Wooden
FENCES

From Henry W. Cleaveland, William Backus, & Samuel D. Backus, *Village and Farm Cottages,* 1856

Wooden FENCES

George Nash

Photographs by
James P. Blair

The Taunton Press

Printed in the United States of America
10 9 8 7 6 5 4 3 2 1

A FINE HOMEBUILDING Book

FINE HOMEBUILDING® is a trademark of The Taunton Press, Inc.,
registered in the U.S. Patent and Trademark Office.

The Taunton Press, Inc., 63 South Main Street, PO Box 5506,
Newtown, CT 06470-5506
e-mail: tp@taunton.com

Library of Congress Cataloging-in-Publication Data

Nash, George.
    Wooden fences / George Nash ; photography by James P. Blair.
        p.   cm.
    "Fine homebuilding book."
    Includes bibliographical references and index.
    ISBN 1-56158-151-8
    1. Fences — Design and construction.   2. Building, Wooden.
I. Title.
TH4965.N37   1997
631.2'7 — dc21   97-18619
                                            CIP

To my dearest and

best beloved, Janey,

in love and gratitude;

for never keeping

"the wall between us

as we go"

ACKNOWLEDGMENTS

First and foremost, I want to thank my editor, Julie Trelstad. Faced with an inexorable production deadline and an incomplete manuscript, she goaded and prodded me until it was done. Thank you, my dear "wicked witch" of the red pen, for your clarity of vision that helped me shape the book, for your persistence and good humor, and for your being the "fist of iron in a velvet glove" that kept me on course.

To Peter Chapman, for meeting the challenge of folding a 3-yd. draft into a 2-yd. book so that nobody would notice; to Karen Liljedahl, for trying to make sense of my scribbled sketches and sorting through mounds of color transparencies; to Bob La Pointe, for blessing the rough art with that archetypal *Fine Homebuilding* style; and to all the rest of you at The Taunton Press for your hospitality, courtesy, and passionate commitment to the high production values so rare in this niche of the publishing world.

And thanks as well and as much to Jim Blair, photographer, for the way you make your camera "see" with the mind's eye instead of taking pictures. Thank you for the special beauty you've given this book and for awakening me to the touch of light upon creation.

To Anne K. Masury, of the Strawbery Banke Museum of Portsmouth, New Hampshire, not only for her valuable assistance in opening the facilities and the archives of the museum to me but also for her generosity in letting me have a copy of her unpublished master's thesis, *Portsmouth Fence Styles*, which was a treasure trove of historical background information. I am deeply in your debt.

To Ed Montague of Aqua-Field Publishing, for turning me on to Oregon photographers Bernard Levine and Bill Rooney, whose images so adroitly extend the regional reach of the text beyond the pickets and pales of New England. And to Charles Miller, for the same.

To Huck DeVenzio of the Hickson Corporation for providing background information on CCA-treated wood and yet more photographs.

To Ken McClelland of the Western Red Cedar Lumber Association and Pam Allsebrook of the California Redwood Association, and to all of the generous people who let us photograph their fences.

CONTENTS

INTRODUCTION

When my wife and I bought our quaintly down-at-the-heels mid-19th century Cape some years back, it was screened from the driveway by a graceless and gateless fence of weathered gray 6-ft.-tall "pickets." The pickets were made from recycled window muntin molding stock nailed to 2x4s spiked to round cedar posts driven into the ground. Although we both liked the idea of a fence that would draw a sharp line between our tended and gardened front lawn and its uncertain dissolution into the gravelly weeds of the driveway margin, we knew it wasn't going to be that fence.

Even if the fence had been well built and standing straight and true, it still would have been all wrong. Seen from the house, 85 ft. back and 6 ft. or 8 ft. lower than the driveway's end, the closely spaced pickets blurred like the spokes of a spinning wheel into a solid screen. A privacy fence was the last thing we needed at the end of a 1,100-ft.-long driveway ending in front of a small house in the back forty of 23 acres of open fields and woods. From our low vantage, it was a great wall across the morning sky. And what message did its style and condition give to the visitor? Or indicate about the

people who lived behind it? Better to live with no fence at all. So I tore it down, with the idea of building a better fence, "someday."

Two years later, when my editor suggested that the book about wooden fences that I was writing needed some photos of a real-live fence being built, and helpfully offered to reimburse the cost of materials, it was an offer I couldn't refuse.

But what kind of fence to build? We knew that we wanted a fence that would be historically and architecturally appropriate and also feel welcoming. This would clearly be a picket fence. Waist-high pickets invite easy conversation across the fence while maintaining the polite reserve that to me is the essence of New England neighborliness, so I opted for a friendly 3-ft. height.

Yes, but what kind of pickets? Delightful and appealing as they may be, the baroque "eye candy" of high-style Victorian picket fences would have been embarrassingly out of place in front of our homespun hodgepodge. But a low picket fence can also appear squat and misproportioned if its pickets are too wide for their height. Narrowing the pickets would counterbalance the lack of height and save the fence from gracelessness.

These are just the kinds of decisions you'll be faced with when designing and building your own fence. This book will help you decide what kind of fence is suitable for your property and guide you through the process of design, material selection, layout, and construction. It will also give you some fascinating background information about the evolution of fences through the ages.

Although I had a great time building my particular fence, I certainly don't mean to imply that it's anything more than a fairly ordinary and pretty well-built fence. If you can swing a hammer, use a saw, and read a tape measure and level, you can probably build a perfectly adequate and maybe even handsome-looking wooden fence. As structures per se, most fences are simple affairs. But there's a lot more to a fence than its parts and the way they get put together.

I am reminded of the high stockade fence that has divided my parents' backyard from their neighbors' for more than four decades. Like some Great Wall of Suburbia, those tight, pointed cedar palings marked the northern border between the world of my childhood and enemy territory. In thinking about that fence, I was struck by the realization that to this day, I don't know the name of the people who lived behind us, how many of them there were, or what they looked like. Talk about your privacy fence! It's almost as if that fence generated a psychological force field even stronger than any merely physical barrier. Which, was, according to Mom, exactly what my folks had in mind when they built the thing. It seems that, in response to our neighbors' complaints about my and my little brother's frequent (and, in their view, provocative) incursions across the line in pursuit of escaped balls, my parents built the mother of all spite fences.

Perhaps more than any other single element of a home's facade or grounds, a fence is as symbolic as it is functional; it not only organizes and shapes the physical space of the domestic landscape, but it also mediates between the public and private realms. It's the line we draw to separate "mine" from "not mine," and "this" from "that." A fence has always (and still does) announced a homeowner's social position and regard for fellow citizens. Whether the message is "keep out" or "do come in," whether it's "move along" or "tarry a while," a fence, by its very nature, cannot help but be significant. Such potent visual impact suggests that any fence ought to be carefully designed. The moral is and always will be: Little things count.

one

THE EVOLUTION
OF THE AMERICAN FENCE

The first man who, having fenced in a piece

of land, said, "This is mine," and found

people naive enough to believe him, that man

was the true founder of civil society.

Jean-Jacques Rousseau, *Discourse on the Origins of Inequality* (1755)

Wood was so plentiful along the Atlantic seaboard that the early settlers had no qualms about using lots of it in their fences.

Long before they began to raise crops or herd cattle, our ancestors had learned to throw up crude fences of tangled branches for protection against wild animals and hostile strangers. Cave dwellers who inhabited a particular site for many generations doubtless built more permanent enclosures of piled stones, or, perhaps, even drove sharpened stakes into the ground. But, since hunter-gatherers and nomadic herdsmen had no use for permanent fences, it wasn't until people began living in fixed settlements built on a foundation of crop and livestock farming that fences evolved to become an important part of the physical and social landscape.

Fences, walls of piled rocks or mud brick, hedges, or, at the very least, a line of planted trees could also be used to mark boundaries and screen the intimacies of family life from the public street. And, wherever some members of society grew wealthier than others, they inevitably built high masonry walls or solid wood fences to protect and insulate themselves from, and announce their status to, their inferiors. Of course, a villa, village, or town that held a store of food and other treasures was an attractive prize to any opportunistic marauders who might be passing through the neighborhood. The palisade, the dense thorn hedge, the moated castle, and the walled city show that fences have always had a defensive function as well. This is reflected in the etymology of the word *fence* itself, which comes to us from the Medieval French *defence*, whose origin is in the Latin "defendere," to knock or strike down, hence, to protect.

The first fences were built from whatever materials were readily available that required little or no tooling. Fences made of stones cleared from treeless fields, sun-baked mud blocks, ditches supplanted with hedgerows of thorny shrubs, saplings, and brush interwoven into tight barriers, and trees hacked into logs and piled upon each other were used by ancient civilizations dating back to prehistory.

Modern fencing was already well established in classical Rome. Writing about 50 B.C., Marcus Terentius Varro (116–27 B.C.), a prototypical gentleman farmer, described in his treatise, *On Agriculture*, a fence remarkably like the same post-and-rail fence of our present-day suburban and village lawns—"built either of stakes planted close and inter-twined with brush; or of thick posts with holes bored through, having rails, usually two or three to the panel, thrust into the openings."

Unlike the Roman estate system, which was built on large landholdings worked by slave labor, from the 10th century until just before the colonization of the New World, English peasants raised their cereal crops and grazed their livestock on largely unfenced common fields and pastures. Under the ancient custom embodied in English common law, livestock owners were held legally responsible for any damage their animals might do by trespass, yet no one was actually required to fence their animals in. Instead, at dawn the village herdsmen would drive cattle from their owners' barns to a common pasture where they foraged under careful watch and then conduct them back to their barns at sundown. Wood was much too valuable as timber for house- and ship-building to be squandered on building fences.

COLONIAL FENCES

Recent historical research has established that, contrary to the popular image, when the first Colonists landed on the shores of New England and Virginia in the early years of the 17th century, they did not immediately confront an impenetrable wall of primeval forest that had to be cleared before they could plant their crops. If that had been the case, the Colonies would not very likely have survived their first year. What they found instead were small fields, only recently abandoned by the Indian farmers whose townsites had been decimated by the smallpox epidemics introduced through earlier contact with European fishermen and fur traders. The first Colonists also found miles of coastal lowlands and estuaries covered with natural pastures of salt grass that could provide forage and hay for their stock.

It was only natural that these newcomers laid out their settlements according to the traditional system of English land tenure. Since, at the time of their arrival, the enclosure of the English countryside was not yet complete, the village common system was, in fact, the only tried-and-true model of land use they knew. Each "proprietor" was given a share (allotment) of the "great lott" (the communal field and pasture) in proportion to the value of his subscription, the size of his family, and his stature in the enterprise.

As in the English countryside, the settlement was laid out with its house lots clustered around the church and village green, which lay at some distance from the common land. Each lot was large enough for a house and its adjoining barn, stables, and kitchen garden. This arrangement also provided some measure of security against Indian attacks, especially when the entire village was surrounded by a palisade fence of closely spaced sharpened logs driven into the ground, as was done by

Some native Americans relied on palisade fences made of sharpened logs driven into the ground for protection against enemy tribes, as shown in this early-17th-century engraving.

the town of Milford, Connecticut, in 1645, or by logs simply stacked up into a high wall, as was done in other towns.

Between building these palisades and throwing up crude shelters for themselves and their animals, the newcomers had so much work to do merely to secure their toehold that only a few able-bodied men could be spared to tend only the most valuable cattle, such as milk cows, sheep, and oxen. The other livestock were let out to graze the salt-hay meadows, and the troublesome and destructive swine were turned out to root through the open woods as far from the village as possible. The open fields used for planting were fairly small and were easily enclosed with crude fences of heaped brush and deadwood or wickets of stakes driven into the ground with saplings interwoven between them, a task the native inhabitants, who kept no cattle or domestic livestock (other than dogs), generally avoided.

Since mixed stock and grain farming and the permanent structures it required were the cornerstone of the Colonists' livelihood, it's not surprising that they saw the Indians' temporary fields and seasonal foragings as proof of their native indolence and savagery rather than as the sophisticated system of food production it really was. The fact that the Colonists' legal theory of land ownership was derived from the biblical injunction to "fill the earth and subdue it" also explains why the Colonial courts were pleased to recognize the Indians' claim to those few lands

that they had actually "improved" by clearing and planting, but not to any of their "unimproved" hunting, fishing, and foraging grounds. As John Winthrop, the first governor of the Massachusetts Bay Colony neatly explained, since the "natives in New England inclose no land, neither have any settled habitation, nor any tame cattle to improve land by,...if we leave them sufficient for their use, we may lawfully take the rest, there being more than enough for them and us."

Conflict between such radically different world views was inevitable and almost immediate. However convenient it may have been for the Colonists to let their livestock forage what was, after all, land free for the taking, their free-ranging cattle wrought undeniable havoc on the unfenced corn and squash plantations of their surviving Indian neighbors. The records of Colonial courts are replete with complaints from Indians over damage done to their crops, clam banks, and other foraging areas by the Colonists' hogs and cattle. For their part, the settlers brought suit against the Indians for injuries to livestock caused by their hunting traps and for animals killed when caught in a native cornfield or pumpkin patch. The Indians petitioned the courts to require the settlers to fence in their cattle. But the courts decided that the Indians had to fence their fields against the Colonists' cattle in order to support any claim for restitution.

The Colonists, in keeping with their penchant for enclosure, built fences of tightly spaced, pointed, split stakes (called pales) driven into the ground and fastened to a horizontal sapling or split rail at their top ends with pegs, nails, or rope binding. The first Colonial record of this pale fence (which is the direct ancestor of the picket fence destined to become a hallmark of the domesticated landscape) is in 1633. The same kind of fence, but on a much larger scale using whole logs instead of poles, was sometimes used to fortify settlements against Indian attack.

SECOND-GENERATION FENCES

Despite their initial setbacks, the American Colonies took root and prospered with remarkable speed. Within the space of barely two decades, most settlements contained gristmills, sawmills, brickyards, and more than a few substantially built and well-furnished homesteads. Agriculture had advanced to the point where food shortages were already a dim memory, and the townships were conducting a brisk trade in forest and agricultural products. One way or another, the Indians in the immediate neighborhood had been "pacified." Even before the first generation of brush and wattle fences that the Colonists had hastily thrown up to signify the intent to enclose a field had rotted away, their builders already had the time and ease to replace them with more durable materials.

Farmers used crossed diagonal stakes and a horizontal rider to lock the corners of worm fences together.

WORM FENCES

The "worm" fence (variously and regionally also known as the Virginia rail fence, crooked rail fence, zigzag fence, snake fence, and rick-rack) may well be a genuine American innovation. In any case, it was certainly one of the first and most ubiquitous second-generation fence forms. Despite its reputed origins in Virginia, the earliest mention of a worm fence is found in the 1652 records of Hampton, Long Island. The worm fence was popular because it was easy to build; in the words of Samuel Deane's *The New England Farmer,* a popular practical handbook published in 1790, a worm fence was "made by lapping the ends of rails or poles on each other, turning alternately to the right and left."

One of the greatest advantages of the worm fence was that it did not require any fence posts. Not only did this contribute to its portability, but it also saved a great deal of grueling labor, both in cutting the posts, mortising the rail holes, and, especially, digging postholes. Given the endemic labor shortage of the Colonial era, the criteria for choosing fence material came down to a question of labor or lumber. Proponents of worm fencing claimed that it used a minimum of labor and a maximum of lumber, which is exactly what the Colonists had during the first century of settlement.

One man, working with an ax and a wedge, could split 150 to 200 11-ft. rails in a day, from sections of white oak or ash logs large enough to yield 30 or 40 rails each. This was enough rail for at least 100 yd. of five-rail fence (the legal minimum; conscientious farmers built eight-rail fences).

Despite its obvious advantages, the worm fence took up space. With corners typically interlocking at about 120°, each zigzagging section of 10-ft. to 11-ft. rails covered about 7 ft. or 8 ft. of ground per panel and ate up 5 acres of land for every square mile enclosed. While flattening the angle of the corners would save space, it also made

Worm fences were easy to build, but they took up space and the enclosed corner sections were difficult to cultivate.

the fence far less stable and more vulnerable to being blown over by strong winds or pushed over by strong cattle. In any case, a cow could easily lift the top rail off the fence panels, and hogs often had to be fitted with a special yoke to keep them from nosing under the bottom rail. To lock the fence together and to secure the top rail,

farmers would drive pairs of rails ("stakes") diagonally across the corners and lay a heavy rail (the "rider") between the crotches of the stakes, as shown in the photo on the facing page.

The worm fence was so universally favored that by 1871, 60% (roughly 4,200,000 miles) of all the fencing in the 39 states that then made up the

Rail fencing anchored by paired, driven posts usually bound together with steel wire
or iron staples made a strong fence that was more economical of materials and
space than worm fencing.

United States was of this type. Only in New England, for some reason, did farmers prefer the traditional mortised post-and-rail-style fence, which had been used since ancient times in Europe. Here, the ends of the rails were tapered to slip through the mortise holes bored through the fence posts, which, soil permitting, were driven deep into the ground. This made a strong, straight fence that used fewer rails than worm fences. Farmers who lacked the time or desire to cut post mortises could even fasten the rails with iron bolts or spikes without an appreciable sacrifice of stability.

BOARD FENCES

Sawmills had been operating through-out the Colonies since at least 1651, and by 1701 board fences were recorded as being used to enclose common fields and graveyards. Since officials in the established settlements were already concerned about the overuse of their town woodlots by then, they had an incentive to choose a fence design that conserved wood for public works. By 1790, sawmill "refuse boards" could be had for $2.00 per 1,000, which was more than enough to build 16 rods of three-board fence. One writer at the time noted that his preference was a fence made of square-cut boards "not much over twenty feet" in length (just try finding boards of that length today), nailed to posts set about 10 ft. apart and at least 30 in. deep (judiciously set 3 ft. to 4 ft. deep in heavy clay or frost-prone soils).

Board fences, which were used to enclose fields as early as 1701, used considerably less wood than worm fences.

DEFORESTATION AND STONE FENCES

If asked to imagine what life was like in the early Colonial settlements, most people will offer some version of men with axes hacking and burning down the towering and dark forests of the "howling wilderness." Even though the northeastern woodlands were almost immeasurably vast by English standards, wood consumption for

The rapid deforestation of the American forest between 1775 and 1825 launched a boom in stone fence building.

heating and cooking, building timber, masts, barrel staves, fencing, and the commercial trade in timber and forest products was so prodigious that the Colonists enacted numerous laws to control it within 50 years after their arrival. For example, by 1689, the town of Malden, in the Bay Colony, had forbidden the cutting for firewood of any tree less than 12 in. in diameter.

Discounting the wasteful clearing practices of the new settlers and the harvesting of wood for home heating and building, the long-term sustainability of the forest cover would have been doubtful in any case. From roughly 1759 until 1865, the manufacturing economy of the United States was almost entirely wood-fueled, either directly or in secondary industrial processes, such as the extraction of tannin from bark for making leather, or reduced to ashes for making potash. Throughout the entire period, much of the eastern forest was cut down faster than it could regenerate itself.

When the clear-cutting frenzy climaxed just before the Civil War, little more than 15% of the original forest cover remained. A good part of those magnificent hardwood forests were burned in the boilers of railroad steam engines. Between 1830 and 1861 over 30,000 miles of railroads were built in the United States (all of which used iron track laid down on hardwood ties). In 1865, Massachusetts railroads burned 53,710 cords, while Vermont's railroad consumed 63,000 cords annually.

The effects of deforestation were already being felt down on the farm long before they reached their apex around 1865. During the Revolutionary War, many miles of fencing had been burned for fuel or destroyed in battle. Farmhouses, too, were often casualties of the fighting. After the war, those farmers who had to rebuild found themselves facing a real shortage of wood for making fence rails.

Although there had been a few isolated pockets of stone fence building as early as the 1640s, particularly where local deposits of flat, easily split sandstone, shale, and schist made wall building easier, stone fencing (including the trademark stone walls of New England) was a relative latecomer to the rural landscape. Most of the 252,539 miles of stone fencing in New England and New York reported in the United States Department of Agriculture (USDA) survey of 1871 were built in the half century between 1775 and 1825, coincidentally, the period of the most rapid deforestation. New England and New York farmers, who had long cursed the native soil for its ability to grow stones easier than wheat, now found themselves in possession of a valuable cash crop. Often all the stone they needed could be found in piles scattered along the margins of their fields where it had accumulated over generations of stone picking.

It's hardly surprising that farmers waited until wood had virtually disappeared before they started building

Crotch-and-rail fences not only saved labor by increasing the effective height of a stone fence but also made the most efficient use of scarce fence rails and posts.

stone fences: Building a stone fence is a laborious endeavor in every sense of the word. The same two men who could put up 200 yd. of worm fence in a day at best might lay 40 ft. of stone fence (but only if the stones has already been gathered and spread along the run of the fence line). At that rate, the quarter-million-plus miles of stone wall mentioned previously represent the labor of 10,000 men for almost 20 years, which is a lot of spare time to find between the other chores of farming. That it was found suggests the relatively high degree of prosperity enjoyed by farmers in the late 18th century. By then, most farmers had become well enough established to have acquired the tools for the job: chain, crowbars, stoneboats, and a team of oxen to pull them. Nevertheless, some farmers saved time and labor by the expedient of building crotch-and-rail fences on top of their low stone walls (see the photo on p. 15).

WESTWARD EXPANSION

The expansion of the railroad network into the Prairie region opened uninterrupted vistas of flat, fertile, and stone-free soil to the plow. New England farmers were leaving the exhausted soil of their rocky hillsides and heading West. In 1850 alone, 168,000 New England farms were abandoned. Since the railroads also made cheap and rapid transportation of food practical, nonmechanized eastern farmers could not compete profitably against midwestern growers. They were faced with the choice of abandoning their farms or of enlarging their fields.

Today, when the closest most of us get to a farm or ranch is the supermarket produce aisle and meat case, it's hard to believe that the economics of farm fencing could have been an issue of nationwide concern in the post–Civil War years, occupying a place in the public forum akin to our present-day debate over the competitiveness of the United States in the global economy. Although the fertility of the Prairie states was well known, the region remained sparsely settled throughout the first half of the 19th century. A major impediment to its agricultural development was the scarcity of trees for building rail fences and a native soil so deep and mellow it grew no field-stones. The invention of balloon framing and the ease of shipping stick framing lumber over the rails helped to under-write the rapid growth of Chicago and other new cities. But the scarcity of (free) native fencing materials and the capital outlay necessary to import them from elsewhere seriously threatened the farmer's profit.

After the Civil War, the U.S. Army, mindful of the strategic role played by fences on the battlefield, appointed General James Brisbin to take an inventory of the nation's fences. A survey included in the Census of 1870 gathered information on fencing in 846 counties, covering almost every state in the Union. The implications of the resulting statistics were eloquently explored in the USDA report of 1871. The conclusion it drew seemed inescapable: The fencing problem was jeopardizing the westward advance of agriculture and, therefore, of civilization itself.

BARBED WIRE

There was, however, one enterprise that did prosper on the Plains. Since the 1850s, ranchers in west Texas had been grazing longhorn cattle on the open range and driving them to railheads in eastern Kansas, where they were slaughtered and shipped to market. Throughout the decade of 1870–1880, western newspapers echoed the fencing debates of the Puritans, with farmers arguing that stockmen should fence the range and cattlemen contending that it was up to the farmers to fence their fields, since they were so much smaller.

Yet events taking place during the winter of 1874 in the kitchen of Joseph F. Glidden's farm in DeKalb, Illinois, were to provide the answer to the fencing problem. About a year earlier, Glidden, attending a county fair,

A major impediment to the agricultural development of the American West was the scarcity of trees for building wooden fences. The invention of barbed wire in the 1870s proved to be the solution. (Photo by George Nash.)

happened to be struck by an exhibition by a Mr. H. M. Rose, which consisted of a 16-ft.-long, 1-in.-square strip of wood studded with sharp nails so that their points were sticking out, intended to be hung on smooth wire, which was fairly common throughout the country at the time. Glidden conceived the idea of placing barbs on a wire instead of driving them into a strip of wood. He found a way to adapt an old coffee mill to twist short lengths of wire into barbs, which he then slipped over a strand of fence.

There were some initial setbacks, such as when thousands of cattle that became trapped against a 40-mile drift fence north of the Canadian River died in the blizzard of 1886. But the demand for barbed wire grew so fast that it soon outstripped production. In response to the shortage, some resourceful ranchers adopted the custom of "lifting" or borrowing the top wire from their neighbor's fences.

Records of the DeKalb factory give some idea of how rapidly barbed wire won the West. In 1875, 300 tons were sold; in 1876, 1,420 tons; and by 1880, sales had soared to 40,250 tons. In 1901, the American Steel and Wire Company produced 248,669 tons. At the same time, the price of the wire dropped continuously, from $20 per 100 lb. in 1874 to $1.80 in 1897.

URBAN FENCES

By the mid to late 18th century, the growth of urban trading centers and the attendant increase in general prosperity resulted in the emergence of the fence as an architectural ornament instead of an agricultural necessity. All throughout the small villages and large towns of Colonial America, the houses of the prosperous came to be almost universally set off from the street by quite elaborate fences. In the thriving trading towns of the Atlantic seaboard, the confluence of a wealthy, erudite, and cosmopolitan merchant class with an abundance of highly skilled joiners and ship's carpenters led to the emergence of a highly developed style of fence building.

Typically painted white, or to match the color of the house, the exquisitely crafted fences sported details that echoed those of the house's exterior trim. Set on granite or brick bases, they featured pedimented posts topped with carved decorative urns, molded and curving rails, turned balusters instead of pickets, and a bedazzling assortment of other "antik ornaments" derived from the classical ideals of beauty and form that were typical of what architectural historians call the Palladian, or Georgian, style (c. 1715–1785).

The inspiration for these fences has been variously ascribed to designs found in the architectural "pattern books" that were imported from England in the latter part of the 18th century. The publication in 1797 of the first edition of the hugely influential *Country Builder's Assistant,* by the American architect Asher Benjamin, completed the nativization of English architectural prototypes. Designs derived from his pattern books were incorporated into the fences and outside trim of the town houses and mansions of the fashionable wealthy. In particular, many of the patterns for the elaborately carved finial urns that cap the gateposts of the Federal-style fences can be found in one or another edition of his book.

In England during this period, iron was the preferred material for ornamental structures. But in America, ornamental cast iron wasn't manufactured in any quantity until the 1860s. Until then, it was imported from England and was consequently so costly that its use was confined to fencing important public buildings and the churchyards of very well-heeled congregations. Whether out of a lingering native frugality, the blush of patriotic fervor, or perhaps simply the impracticality of an excessively long lead time in taking delivery, even notables who could otherwise afford iron apparently preferred to patronize carpenters who worked in wood. Proof that these artisans were capable of

In urban areas,
the fence emerged
as an architectural
ornament instead
of an agricultural
necessity.

Typically painted white, the exquisitely crafted urban fences of the 18th century sported details that echoed idealized classical Greek and Roman architecture.

The small diameter of the rounded dowels and the alternating-height picket arrangements on urban American fences (left) were intended to mimic the profile of their cast-iron English prototypes (below).

building wood fences of a perfection and beauty that rivaled if not surpassed the most elaborate ironwork can be found in the surviving fences and historic re-creations that grace the historic districts of old New England seaports such as Portsmouth, New Hampshire, and Newburyport and Salem, Massachusetts.

NINETEENTH-CENTURY FARM FENCES

In the countryside, until about 1820, the front yards of most farmhouses were left barren and unfenced. The formal front yard, with its manicured lawn and symmetrical groupings of trees, its shrubs, paths, and neat, white-painted board fences and gates, was one manifestation of a conscious campaign of general self-improvement by progressive farmers who yearned to fulfill the Republican ideals of the Greek Revival period (c. 1820–1850) and of the general tenor of the age,

which exhorted its erstwhile frontiersmen to adopt the manners and trappings of gentility.

The Puritans had always regarded the "improvement" of the landscape as a moral imperative. To them, and to their philosophical heirs, strong fences and stone walls were public testimony of sober industriousness and enduring monuments to the beneficence of

"Ha-Ha" Wall

A "ha-ha," or sunk fence, helped contain the livestock without obstructing the view from the house.

Divine Providence. This strain of thought evolved into a notion widely advocated by progressive agriculturists in the early years of the 19th century. Writing at a time when many felt that America was close to actualizing the Jeffersonian ideal of a nation of sturdy yeoman farmers, these commentators asserted that the visual order of the farm gave proof of the moral order of both the public and personal landscape.

Orderly fences became emblematic of personal prosperity and probity.

In the 1840s, the winds of fashion began a shift that lasted through the 1850s. At the very same time the sophisticated country squire was building his picket fence across the front yard, the aesthetic theories of the Romantic movement had crossed the ocean to infect the avant garde with an antipathy toward the very idea of fencing. It became fashionable for landscape architects and their upper-crust clientele to eschew what they

regarded as the sterile formality of the courtyard lawn in favor of the more "natural" theory of blending the house and grounds into the general landscape. The most influential critic of the front-yard fence was the famous landscape architect Andrew Jackson Downing. What he preferred was a stone "sunk fence, or ha-ha" instead—a wall 3½ ft. high, cut into the bank to contain the livestock without obstructing the view from the house across the landscape to the river.

VICTORIAN FENCES

To the eternal consternation of the avant garde, some of whom had rushed to tear down the fences of their country homes, the bulk of society continued to build picket fences. Fashionable or not, the white picket front-yard fence became firmly established throughout the country—not only as an emblem of dignity and charm for the modest home of the genteel urban and suburban middle classes, but also as a poignant badge of civilization in the precarious settlements of the western wilderness.

One immediate effect of Downing's influence was the popularization of the Gothic Revival style that was the immediate predecessor of what, beginning in the 1860s and lasting almost to the end of the century, would become the Victorian style. For many, the precipitous transition from a rural agricultural economy to an urban industrial one was an opportunity to build great fortunes. This frenetic, almost exuberant exploitation of commercial possibilities found its social expression in an unprecedentedly ostentatious public display of wealth.

In such an atmosphere, it's hardly surprising that the already highly articulated detailing of the earlier classic Greek Revival style fences mutated into the fulsome and almost obsessive Victorian preoccupation with decorative elaboration. The introduction of steam-powered industrial bandsaws and scrollsaws and other specialized woodworking machinery, together with the development of a nationwide transportation network, made it possible to mass-produce and ship affordable decorative millwork and fence pickets ordered out of catalogs anywhere throughout the country. Improved foundry techniques brought down the cost of cast iron to where it was within reach of the average wealthy businessman. At the same time, the labor-intensive handwork that had designed and built the older fences became too costly to continue.

This page from Palliser's New Cottage Homes and Details (1887) shows the rich variety of fence patterns available in the Victorian era.

Fences, Gates and Posts.

Although some contemporary critics carped that the mass production of fence pickets would soon lead to a landscape criss-crossed by identical off-the-shelf picket fences, the truth is that the number and variety of patterns available and the permutations into which they could and were combined were almost beyond imagination and certainly anything but boring (see the print on p. 23). Designs ran the gamut from the simple picket terminated with a variously patterned chevron, to intricate "cutwork" designs with exotic names such as Lambs and Weeping Willows, Italian Villa, and Empire Torches. The fence posts were typically massive and topped by oversized, multilayered caps that lacked the carved urn finials of the earlier styles.

Suburban Fences

Many people equate the birth of the suburbs with the birth of the baby boom, but suburbs were already well established by the last quarter of the 19th century. To architects and social critics thinking and writing about this new form of settlement, neither urban nor rural, where houses and their grounds enclosed a private community that faced a public street, the messages conveyed by fencing were an important issue. The social message of a fence, more than its function, per se, was the arbiter of its form.

Thomas Hill, the author of a handbook of business and social etiquette widely used throughout the 1870s on into the 1920s, regarded "division fences between houses" as an atavistic throwback to the high enclosures of a more barbaric age. By his lights, a fenced yard was symptomatic of more than merely disagreeable neighbors; it signified a lack of moral advancement, which had profound civic consequences (see the prints on the facing page).

At the same time, another sensibility was emerging from the social dislocations attending the aftermath of the Civil War, the disillusionment with Reconstruction, and the financial uncertainty in the wake of the Panic of 1873. A discontent was rising, like a choking pall from the smokestacks of the new urban industrial culture. A revulsion against the social, political, and economic stratification, against the noisome factories and the filthy and unhealthy living conditions of the cities, fueled a nostalgia for what many Americans remembered as a stable, simple, and wholesome past. In this light, everything Victorian was seen as garish, excessive, and false. In architecture, this disgust manifested itself in the Colonial Revival movement, which began in the 1890s, peaked in the 1920s, and lingered on until the Great Depression of the 1930s.

The ostentatious wood and cast-iron fences of the Victorians were pulled down and replaced by fences whose clean, white lines hearkened back to the purity of an idealized Colonial past. Writers like Wallace Nutting engraved the mythic image of the quaint New England village, with neat white houses surrounded by neat white picket fences, on the public mind as a preeminent symbol of our national heritage and pride. Following the lead of architectural historian Alfred Hopkins, designers and builders turned to the fences of "ancient" Salem and Newburyport, Massachusetts, for their models. The momentum of the Colonial Revival movement petered out in the Great Depression. Maintaining or rebuilding the front fence was, understandably, assigned a low priority after the Depression. Many were torn down or simply left to fall down by themselves. Of the city fences that did survive, many more were casualties of the wave of urban renewal that swept our cities clean of the detritus of history during the early 1960s.

Since the 1980s, the gentrification of formerly blighted areas and a concern for preserving our architectural heritage, coupled with yet another reemerging cycle of nostalgia for the past, have rekindled an interest in building old-fashioned decorative wood fences. Only this time around, it's the Victorians who, quite appropriately in

FIG. 22. PEOPLE WHO ARE TROUBLED BY THEIR NEIGHBORS.

The social message conveyed by a fence was an important issue for 19th-century commentators, as shown in these two prints from Thomas Hill's handbook on social etiquette.

FIG. 23. THE NEIGHBORHOOD WHERE PEOPLE LIVE IN HARMONY.

*After the Civil War, the ostentatious wood and cast-iron fences of the Victorians
were pulled down and replaced by fences with clean white lines that hearkened back
to the purity of an idealized Colonial past.*

our present age of rampant social and economic stratification, are providing the template.

Across the rest of the American landscape, except in those quaint rural villages, the front fence left for good along with the domestic livestock. The suburban developments of the post–World War II era are notable, among other things, for their lack of boundary fencing. It might seem that the current lack of a formal symbolic barrier in the front yard corresponds to the casual, comfortable open face Americans are wont to present in public; the front yard as the landscaping equivalent of jeans and a T-shirt. But in modern America, the focus of family life is no longer toward the street. Except for a vestigial portico, the front porch has disappeared altogether in both urban and suburban settings; in the former, because the street is now too threatening, and in the latter, because there is no automotive equivalent to engaging in friendly conversation with the casual passerby. The front yard has become a symbolic demilitarized zone, demarcated by, at most, a panel or two of a rail fence that sketches a boundary, implying a protected zone beyond which one should not proceed without invitation. Fencing ordinances implicitly recognize the purely symbolic nature of such fencing by restricting them to waist height.

In modern suburbia, the front yard is typically unfenced, while the backyard is defensibly enclosed. (Photo by Huck deVenzio.)

The backyard has become an extension of the private domain of the house and as such is now often surrounded by high, opaque, secure fences, both for privacy and safety. These fences enclose and protect children, pets, and swimming pools, and, because they are too tall to see over and too solid to see through, they isolate each family from their neighbors, thereby incubating a lack of communal feeling entirely antithetical to the impulse that first created the suburbs.

The role of fences in mediating intercourse between the public and private spheres of domestic life in the last quarter of the 19th century seems remarkably prescient and relevant to the conditions of our own present-day neighborhoods. As our public life, or at least our perception of it, has grown more threatening, privacy and security fences are becoming increasingly

popular throughout the country. The implosion of gated communities can be seen as a disturbing retreat from civility (in its fullest meaning as encompassing both etiquette and public life).

The remarks of Alfred Hopkins are an ironic commentary on the present-day passion for security fences and walls. "Once upon a time, when our ancestors spoke of their 'defences,' they referred to the great walls and battlements which protected them against their warlike neighbors; but, nowadays our neighbors are more neighborly, and the 'defences' have dwindled down to 'fences.' The evolution of the fence has proceeded in accordance with 'the nature of the marauders to be shut out.'"

two

WOODEN FENCE TYPES

Before I built a wall I'd ask to know

What I was walling in or walling out,

And to whom I was like to give offense.

Robert Frost, "Mending Wall" (1914)

*The white picket
fence that divides
the front yard
from the village
street is the classic
boundary fence.*

When you set out to design a fence, the first question you need to ask is "What's it for?" Is it to mark a boundary? To enclose or define a special area? Is it to ensure privacy? To deter unwelcome visitors, or to keep pets or young children secure? Is it to enhance the appearance of your property or landscape plantings? Or to block an unpleasant view or chilling wind? To dampen the sound of traffic? It's likely that your fence is intended for more than one of these purposes or that different sections of the fence will fulfill different needs. Establishing functional priorities is the first step toward intelligent planning. Since such design factors as height, transparency, strength, choice of infill material, and myriad other construction details will differ according to the intended function of the fence, listing the purposes of your fence in the order of their importance can help you choose between conflicting options.

BOUNDARY FENCES

The need to define one's turf is such an unremarkable universal phenomenon that it may well be instinctual. Humans probably build fences for the same reasons that other animals mark their territories with their scent. And like those invisible borders, human boundary fences don't have to be substantial to be effective. In fact, what distinguishes the boundary fence from other fence types is that its effectiveness depends more on the establishment of a psychological barrier than a physical barrier. Such fences are intended to suggest and set limits; they're a kind of metering device that regulates the intercourse between public and private life.

Any fence can be a boundary fence. Indeed, it could be argued that every fence is essentially a boundary fence. The critical design variable is not the message of demarcation itself, but the tone of that message. While a picket fence is friendly, a high solid wall is forbidding, and a barbed-wire-and-chain-link fence is downright hostile. However, the archetypal boundary fence is a polite fence. Its effectiveness derives from respect for the rules of social etiquette rather than from the brute force of an impenetrable wall.

The minimalist post-and-rail "drift" fence draws the line between the ranges of suburban ranch houses.

This boundary fence of sharp iron spears could be off-putting, but the otherwise harsh message is sweetened by the decorative filigreed infill.

A very low fence, like this one bordering a Maine graveyard, directs traffic and creates a symbolic barrier line, but it does not prevent access to the fenced area.

A boundary fence can define areas within a yard for specific purposes, such as to separate a garden from the lawn, to enclose a pool, or simply to create a nook or refuge apart from the larger yard.

This trellis-cum-fence creates a refined and formal entry that both prevents pedestrians from taking a shortcut across the lawn and showcases ornamental plantings.

As such, it can exist meaningfully only in a civilized society, one in which everyone agrees that "good fences make good neighbors."

A fence that is only a symbolic barrier is most often a low fence (certainly no more than 3 ft. 6 in. high). It can be nothing more than a line of painted rocks or small evergreens. It should present a tangible enough barrier that no one would be tempted to squeeze through it without guilt, yet be open enough not to block the view behind it completely. Rail fences, horizontal board fences, and picket fences are the obvious candidates.

Besides marking the perimeter of your property, boundary fences can also shape and define spaces and patterns of use within it or function as traffic-control devices. Not all separations need to be prohibitive. An attractive low fence can also invite the eye of a passing stroller to linger on the view or direct the visitor toward a landscape feature.

The modern stockade fence, with its vertical half-round sharpened palings,
is a scaled-down version of the palisade wall that Colonists built of stout, closely
spaced logs driven deep into the ground and pointed at the tops to protect early
American settlements.

Security Fences

While the need to mark territory may be instinctive, it isn't essential to survival. Defending yourself against enemies or predators is. Long before there were yards and streets, people were building defenses to keep wild beasts and strangers out of their shelters and away from their livestock. The modern stockade fence is a relative of the palisade wall.

Someone intending to break into your house is already too far beyond the pale of social convention to be stopped by a picket fence. Nevertheless, even though it won't keep intruders out, a picket fence will keep small children and the family dog from wandering out into the traffic or falling into the swimming pool. Until recently, in most parts of the country, this was all the security homeowners felt they needed. Since, in most communities, zoning regulations still restrict the height of fences in the front yard to less than 3 ft. 6 in., in an age when parents worry how to keep intruders from invading the home ground, children and dogs play in the back yard, protected by a high, strong fence.

Since the purpose of a security fence is to keep something out or something in, impregnability is its most salient design element: An effective security fence should be tall, strong, and hard to

The picket fence traces its ancestry back to the sharpened pales of the palisade wall. Pickets can be a bit tricky to hop over, but their only actual deterrent effect on human intruders is psychological.

*The sawtooth tops
of this California
security fence dis-
courage climbing.
(Photo by Bernard
Levine.)*

*The smooth,
tightly spaced
boards and the
lack of intermedi-
ate rails make
this an effective
security fence.*

climb over. A tall wood fence of closely spaced pointed pales or stakes will usually discourage all but the most determined intruder from climbing it. However the horizontal rails (or "stringers") to which the vertical pales or boards are fastened can make a convenient ladder for climbing the fence. The obvious solution is to build the fence so that its rails face inward, leaving the smooth face to confront possible intruders. This can be a problem in two ways: First, it provides an escape ladder for the children the fence is supposed to keep confined. (Although typical 2x4 stringers installed flatwise not only provide a narrower toehold but also add to the rigidity of the support.) Second, you have to look at the "back" side of the fence unless you build a "good neighbor" fence which is boarded on both sides of the stringers.

Fence boards don't need to be vertical to discourage climbers. Diagonal boards spaced closely enough to prevent a toehold (no more than an inch apart) or laid at an angle greater than 45° are hard to climb. Horizontal boards spaced less than a toehold apart also discourage climbers, especially if they overlap clapboard-style or join together in a solid tongue-and-groove panel. Basketweave-style boarding provides as many handholds and

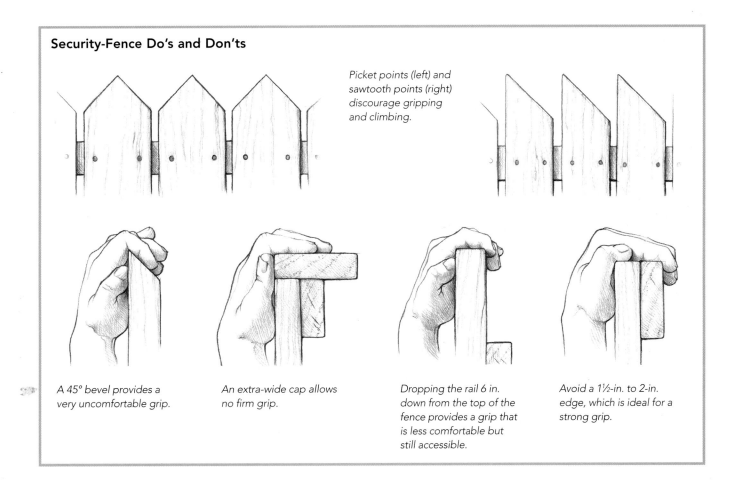

Security-Fence Do's and Don'ts

Picket points (left) and sawtooth points (right) discourage gripping and climbing.

A 45° bevel provides a very uncomfortable grip.

An extra-wide cap allows no firm grip.

Dropping the rail 6 in. down from the top of the fence provides a grip that is less comfortable but still accessible.

Avoid a 1½-in. to 2-in. edge, which is ideal for a strong grip.

The utilitarian appearance of a chain-link fence can be mitigated somewhat by weaving wood or PVC slats into the fabric. (Photo by George Nash.)

toeholds as a jungle gym and shouldn't be used if security is more important than privacy.

Although security fences must be strong and tall, opacity is not an inherent design requirement. As long as the infill material is spaced closely enough together that people or animals can't squeeze through, the fence itself can be open enough to allow whatever amount of view through it you want. For example, an infill of tempered glass or clear plastic panels would be impossible to climb over, yet would allow completely unrestricted vision.

In this sense, the chain-link fence topped with barbed or razor wire is the paradigm of the most effective, albeit egregiously ugly, security fence. The needle-sharp pales of ornamental

wrought-iron fencing make a much more elegant and equally off-putting barrier where appearance is more important than utility and cost is no object. Because it so clearly announced its owner's wealth while maintaining a discreet but impregnable defense, ornamental ironwork was the security fencing of choice for the town houses of the Victorian gentry. When privacy was as important as security, their estates were traditionally surrounded by an equally effective barrier of stone or brick wall topped with embedded broken glass, sharp stones, or iron spikes.

Even if you could afford to, you probably wouldn't be allowed to install any of these types of high-security fences around your home. Most zoning regulations prohibit the use of "hazardous" fence materials, that is, those that are intended to cut, impale, electrocute, or otherwise cause bodily harm, in residential areas. Except for wrought iron (or its less expensive modern equivalent, tubular steel), which is usually not considered a hazardous fencing material, these types of fences are restricted to commercial, industrial, or farm use.

Of course, many homeowners do fence their property with chain-link fencing, sans barbed-wire extensions. It certainly won't increase your property value as much as a wooden fence will, but chain-link fencing is one of the best choices for utilitarian high-security enclosures within your property. There's certainly nothing better for a dog run or a tennis court. Chain-link fencing is also excellent for protecting a swimming pool.

A security fence of wire mesh (either galvanized or plastic-coated steel) suspended between wood stringers doesn't have the industrial look of chain-link fencing. So long as height (standard mesh varies from 26 in. to 55 in. wide) and extra strength aren't critical design requirements, wire mesh makes an adequate security fence, especially for internal enclosures. It's particularly good for enclosing children's play areas since, unlike a board or picket fence, the light-gauge mesh is almost visually transparent, so parents can have a clear view of their children while they play. The closely spaced wires are also excellent for keeping pets and (nonclimbing) animals out of gardens.

PRIVACY FENCES

Throughout the second half of the 19th century and up until after World War I, the front porch was the nexus between family life and village life. In the ideal suburbs envisioned by landscape architects and city planners of the time, there would be no side fences between lots. The lawns would run together, to make a semiprivate park, which would be discreetly separated from the public street by a nearly transparent iron fence. The backyards became the locus of family life. Because children and pets didn't

The primary purpose of a privacy fence is to provide a visually secure area. (Photo by Bernard Levine.)

A privacy fence with solid infill can be as ornamental as a more open fence. The softly rounded panels of this fence are friendly yet still effectively block the view.

private. A desire for privacy is as natural a reason for building a fence as a desire for security. Indeed, the two purposes are often inseparable from each other.

The most salient aspect of a privacy fence is opacity: The most effective privacy fence is one you can't see over, under, around, or through. The degree of opacity depends on how much privacy you need. Wooden privacy fences most often feature infill panels of closely spaced, overlapped, or tightly butted boards. Plywood panels also make a very effective privacy screen.

The major difference between a privacy fence and a security fence is that the former is intended to screen activities within the enclosure from the view of those outside it, whereas the intention of a security fence could be said to screen the presence of outsiders from the insiders. The distinction is somewhat arbitrary, of course, since the visual screening effect of a privacy fence depends not only on your point of view looking out but also on the point of view of someone looking in. The distance and the elevation of the view to be screened out or framed in relative to the observer's line of sight are factors to consider when calculating fence height (see pp. 69-70).

Just as a symbolic boundary can act as a real barrier, a more open slat or lattice fence can still suggest a feeling of privacy. The eye tends to stop at the fence rather than look through it. Another way to provide privacy without

necessarily respect the property rights of the neighbors, these areas were almost universally enclosed by some sort of secure fence. Although the surrender of suburban streets to the automobile has made polite conversation between a passerby and a householder impractical, the pattern of open or only symbolically fenced front yards and securely fenced backyards is still almost universal in the residential areas of urban and suburban North America.

But sometimes even though a structurally secure fence may not be called for, a visually secure fence may be. Sometimes, we just don't want to worry about what the neighbors might think. Most people feel that sunbathing, swimming, outdoor dining, and many other family activities should be

A high wooden fence with solid panels provides maximum privacy, blocking both sight and sound. (Photo by Bill Rooney.)

Any fence above eye height can function as a privacy fence. Even without solid infill, this fence offers a visually interesting arrangement that focuses the eye on the fence rather than on what lies beyond it.

Varying the height of a privacy fence can counteract the feeling of claustrophobia that can result when a solid fence encloses a small area. (Photo by Bernard Levine.)

erecting a solid wall is to use lower panels to enclose less-sensitive areas. Or you might mix a low fence with tall evergreens, running the fence in front of the planting or pruning the lower branches so that the fence fits beneath them. A somewhat open fence can also counteract this feeling of confinement while still maintaining the illusion of privacy. A vertical board or louvered fence with spaced boards alternating between opposite sides of the stringer is a good example of this principle (see the photos on the facing page).

Because the solid infill of a privacy fence can present a rather bland or even uninviting facade to the street, designers vary the rhythm or texture of the boarding to counteract this tendency. Possible alternatives include running boards of different widths on the same or opposite sides, alternating infill panels on opposite sides of the fence posts, or varying the surface texture by using board-and-batten siding, bevel siding, or even wood shingles. Decorative cuts at the tops of the fence boards can also enliven the privacy fence.

SCREENING FENCES

Functionally speaking, "screen" and "fence" are almost synonyms. Defined as a partition that separates, conceals, shelters, shields, or protects, the word *screen* suggests solidity, or at least the opacity of a privacy fence. Interestingly enough, its secondary meaning as a kind of sieve also correlates with

When a vertical louvered fence is viewed straight on, the fence appears to be solid (above), but when it is seen at an angle, the alternating gaps permit a limited view through the fence (right).

A solid fence can be open to the breeze to provide a screen against views, wind, and bright sunlight. (Photo by Bernard Levine.)

another important functional aspect of fencing. Just as a window screen permits a view and a breeze, while keeping insects out of the bedroom, fences can be used to moderate the local environment. While opacity may be desirable for some screening purposes, others may call for a greater degree of transparency, depending on whether the need is for visual, aural, or climatic buffering. Thus, the salient aspect of a screening fence is controlled permeability.

For example, a fence that's intended to block wind wouldn't necessarily be equally effective for screening traffic noise or camouflaging undesirable features. Counter to what might seem logical, a solid barrier actually doesn't provide very much protection against wind. The wind simply blows up over the fence and continues on with undiminished velocity across the yard. Although this kind of fence does create a narrow zone of calm air on the downwind side about equal in width to the height of the fence, this dampening effect is sometimes counteracted by downdrafts that arise when the encounter between a gusting wind and the fence is too abrupt. The low pressure in the eddy formed on the lee side of such a fence actually causes the wind blowing over the top to plummet down to ground level.

A 45° baffle that angles away from the wind to the top of the fence extends the protective zone somewhat further. If the baffle faces into the wind, it will spread the attenuation effect over an even larger area, to about twice the fence height. But, compared to solid infill, even better results are possible with any infill that has small to medium-size openings, such as lattice, slats, lath, louvers, or basketweave. These tend to break a steady windstream into turbulent eddies. Wind-tunnel tests have proved that down-canted horizontal louvers are one of the most effective designs. Although the calming effect is less pronounced immediately downwind of the fence, it's significantly greater in the region between one and two times fence height. As the drawing at right shows, softening the breeze is felt as a rise in relative air temperature.

Any fence design that softens the wind will also tend to diffuse driven snow or rain. Snow tends to pile up against the upwind side of a solid fence. This would be beneficial if a driveway were on the downwind side of the fence and troublesome if it were on the windward side. Another possible drawback to using solid fencing to control snowdrifts is that the pressure of the snow piled up against the fence could eventually push it over.

Fences can be designed to modulate sunlight, controlling its intensity from full shade to full sun, and to reduce heat and glare. For example, a solid board or masonry fence oriented east to west will be in full shade on its north side and in full sun on the south. Horticulturists in 17th-century France and England took advantage of the solar heat stored by a south-facing wall

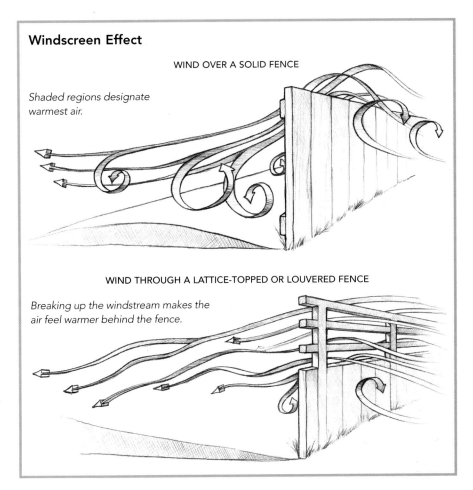

Windscreen Effect

WIND OVER A SOLID FENCE

Shaded regions designate warmest air.

WIND THROUGH A LATTICE-TOPPED OR LOUVERED FENCE

Breaking up the windstream makes the air feel warmer behind the fence.

or fence to harvest warm-weather fruits from espaliered trees and vines. Since a fence painted white will reflect sunlight, many gardeners in temperate areas increase their tomato crop by planting their vines in front of just such a fence. An ill-considered fence could also just as effectively shorten the growing season for vegetables planted in its shadow.

Whatever its limitation as a windscreen, a solid fence does make the most effective noise barrier. In fact, the higher and thicker the fence, the better it will mute sound. The ideal sound screen has a smooth, unbroken face, which tends to reflect sound away from the area to be protected. Although a masonry fence is the best solution for noise reduction, a high wooden fence, sheathed on both sides

A lattice fence acts as a screen to soften the wind and to modulate the brightness of the sun to improve garden growing conditions.

A high solid fence, in addition to providing good security and privacy, is the most effective barrier against noise.

with solid infill, also makes a fairly effective sound barrier, so long as the joints between the infill boards are sealed—as, for example, with an overlapping board-on-board pattern or a traditional board-and-batten style. A plywood-panel infill or a wood-shingled or clapboard-wall type of fence also offers a fair degree of sound attenuation. (Here, the rough texture of the wood tends to diffuse and absorb the sound waves, instead of reflecting them.) This same diffusion/absorption effect of densely textured materials is the reason why living hedge fences are often used to buffer the noise from heavily traveled roads.

In most cases, the practical or legal limits to fence height and thickness will preclude the construction of a sound barrier whose effectiveness is derived solely from its physical properties, particularly since it's unlikely that the fence will be high enough to block the offending noise. As is the case with privacy and boundary fences, sound buffering also has a strong psychological component. Curiously, if you can't easily see the source of a noise, the noise is less noticeable. In effect, buffering your perception of the noise often has the same soothing effect as attenuating the noise itself. Thus if only a narrow strip of yard separates a house from a busy street, a high, tight stockade fence would maintain a pleasant ambiance behind the fence line.

There are some aspects of visual screening that fall outside the functional domain of the privacy fence proper. For example, fences can hide garbage cans, compost piles, lawn tools, piles of "surplus" materials, and other visually unappealing features of a property. In some communities, local zoning laws may require any RVs stored on the property to be camouflaged by a fence.

ORNAMENTAL FENCES

Whatever their functional purpose, fences are also architecture, or ornament. At the very least, a fence can increase or lower property values. At the very best or worst, it can add to or detract from the beauty of a house. To an architect, the most significant aspect

The strong vertical and horizontal plane defined by a boundary fence visually extends the line of the property, expanding its apparent size.

Ornamental fences add balance to the relationship between a house and its surrounds.

A decorative fence can provide architectural embellishment for the house and the yard as much as the cornice trim does for the house and the roof.

of a fence might be that it defines a strong vertical plane. Think of a fence not as a two-dimensional line marking off sections and borders on the flat grid of your site plan but as a solid slab enclosing and subdividing the volume of space surrounding your house. In this sense, the plane of a fence acquires a dynamic. The use of border fencing to shape or define space has already been remarked upon. Consider how the implied planes of the fences placed on either side of a small house visually extend its mass toward the street, expanding its apparent size; or how, when a house is relatively close to the street, a low fence can make its small front yard appear bigger; or conversely, how, when a house is floating on a large lot, surrounding it with a fence seems to anchor it to its site. In this way, fences can be understood to bestow stability and add balance to the relationship between a house and its surrounds.

One way a fence can accomplish this connection is through the repetition of rhythmic elements or the use of symmetry. (The language of visual design is unavoidably metaphorical, but it still says something worthwhile.) Instead of a slab, a fence can be a blank canvas or a picture frame: Because they are repetitive, the lines of the fence infill and the metronomic intervals between boards or rails or between the posts count out a visual rhythm. The location and width of gates and infill panels and their juxtaposition to the rhythmic elements of the house (such as

Even a rustic fence can exhibit architectural balance through rhythmic repetition of its elements. (Photo by Vincent Lawrence.)

The elements of the fence are like notes on a scale or brushstrokes on a canvas; the texture, color, and detailing of the infill become critical parts of the composition.

Echoing details of the exterior trim such as moldings, decorative filigree, or finials creates the effect of an architectural connection between the house and the fence.

This gate sets off the entry without breaking the flow of the fence.

its doors and windows) can establish pleasing symmetries.

As was mentioned previously, a fence can also be understood as a kind of diffusion membrane. By establishing the boundary between the public street and the private house it defines the front yard as a semiprivate transition zone. This welcome zone of calm eases the stress of the passage from the street to the home. As a kind of outdoor vestibule, the fence is also an extension of the house. As such, it's important that the style of the fence and its

materials and finishes have some obvious connection to those of the house proper.

Gates are an antic element and the locus of elaboration in the scheme of most fences. A gate whose style deliberately contrasts with the overall pattern of the fence will capture the visitor's attention, leading them to it. Other gates are as unobtrusive as possible, blending so thoroughly into the flow of the fence that they virtually disappear, thereby preserving the unity of the design.

Like a trellis or an arbor, a simple wooden fence can be used as a backdrop or support for ornamental plantings.

Besides the gate proper, the posts that support the infill panels are often used as a decorative embellishment. If any part of the fence will duplicate the details of the house trim, it's most often the gateposts. The design can range from a basic square-edged post given a bevel-cut or chamfered top to a post ornamented with built-up layers of trim, raised, fluted, or chamfered panels, molded and dentiled caps and bases, or elaborate finials.

FENCES AS AN ELEMENT IN THE LANDSCAPE

Of course, a fence is more than just architecture, more than a plane in space that defines volumes or a pleasing composition in itself. It's also embedded in the landscape; it is landscape. Can you picture a rural landscape without a rambling rail fence or a tumble-down stone wall? The connection between fences and plants always seems quite natural. (If you want to see just how

Carefully selected plants can soften the facade of a stark fence and add color and depth to it. Annual flowers ease a newly built fence into the landscape, as if it had always been there.

Just as a fence eases the transition between the public and private realms, yours can join the natural to the architectural by combining design themes and finish details from both realms—architectural order inviting a friendly vegetable embrace.

natural it can be, stop weeding your fence line for a season.)

When you choose the type or color of the fence treatment, consider how the fence will harmonize or contrast with the foliage or blooms of your plantings and vice versa. There's something perfect about the brazen splash of red roses against the brilliant white slats of a picket fence. Like the walls of a gallery, your fence can showcase specimen plants yet remain prominent and distinct from them. Or, festooned with rambling vines, it can discreetly merge with the landscape.

There's one caveat when plants and fences join: Dense mats of vines and heavy foliage plantings against solid infill block drying air movement and provide a cool, moist, and shady enclave in which algae and moss can flourish. When subjected to constant high moisture and corrosive secretions from its vegetative coating, any protective paint film soon blisters and flakes away, exposing the underlying wood to attack by wood-destroying fungi. That is why garden fences often feature an open type of infill design or one that minimizes the kind of details that trap and hold water to offer greater longevity.

Because garden fences are often covered with vegetation, which means a constant exposure to moisture, open designs will best resist premature decay.

three

DESIGN FACTORS

To unite strength, beauty, and "transparency,"

is the object to be gained. What

wooden fences will best do this,

we must leave to the reader's ingenuity

and good sense to decide.

Frank J. Scott, *The Art of Beautifying Suburban Home Grounds* (1886)

When you set out to design a fence, you have, on one hand, the endless possibilities of your imagination and, on the other, the limitations of your wallet. In this chapter, I'll outline both the aesthetic possibilities and the real-world limitations—such as your view, uneven terrain, and fence laws—that will all contribute to determining the form of your fence.

AESTHETICS AND STYLE

It's one thing for a fence to create what 19th-century landscape architects considered a realm of "taste and gentility" and quite another to decide exactly what a tasteful fence should look like. Defining matters of aesthetics and style is a tricky business, but there are some important general aesthetic considerations that are particularly applicable to wooden-fence design, such as openness, formality, balance, contrast, and scale.

OPENNESS

In the broadest sense, all fences can be understood as fitting into one of two stylistic categories, "open" or "closed." These mean exactly what they say: Open fences have spaces between the infill units; closed fences have solid

The rail fence is a classic example of an open fence.

A closed fence is one that has no spaces between its infill boards.

infill. Rail and picket fences are open, whereas palisade, grapestake, and most board fences are closed. And some fences, like basketweave and louvered styles, are a little of both. This classification is based on visual, not structural, quality; that is, how much you can see through the fence rather than how easy it is to climb over. For example, even though it presents a solid face, a glass fence is still an open fence.

FORMALITY

A second general distinction can be drawn between "formal" and "informal" fence styles. Elements of fence style that suggest formality typically include the use of smooth-planed and painted (usually white) wood, classical decorative detailing of

With classical decorative detailing and a strong emphasis on symmetry, the fence above presents a very formal facade.

The low pickets and log posts and rails make a very informal fence, well suited to the day-care playground it encloses. (Photo by George Nash.)

The visual tension created by the strong contrast between the vertical pickets and horizontal clapboards is heightened by their close proximity.

fence and gate posts, and a strong emphasis on symmetry. Although for the most part formal can be considered synonymous with "traditional," the two aren't necessarily quite the same. For example, a low white picket fence is always traditional, but it can be rather informal if the pickets themselves are plain square-edged slats coupled with

unobtrusive and unadorned posts. On the other hand, it can be quite formal, as, for example, when a balustrade of squared pales is paired with prominent and highly decorated posts.

Regularity is also an aspect of formality. Think of the geometrically ordered arrangement of a formal English garden. The deliberate repetition of design elements creates similar patterns in a fence. So does a central gate opening onto a neatly paved path leading directly to the front door of the house. Although most people automatically associate formality with rectilinearity, an entrance flanked on both sides by a curving fence line replicates the extreme formality of a Palladian estate.

Rough texture and a naturally weathered or dark-stained finish are two of the most significant elements of the informal style. Although the close and ubiquitous association of Northwest Coast–style cedar and redwood board fencing with contemporary house styles probably defines the informal fence style, some very traditional fences are also definitively informal. Split- or peeled-pole cedar rail fences are one example. Grapestakes (rough, rectangular slats of split redwood) and palings are among the oldest types of infill, and a fence built with them could never be considered anything but informal, whether paired with an antique or modern house.

Formality does not have to mean rectilinearity. Here, a curving fence line replicates the formality of the estate beyond.

This odd little fence seems grander than it actually is because it fits visually with the style, details, and colors of the house it fronts.

The unobtrusive pickets and prominent gate-posts seem to frame the house, emphasizing its beauty rather than calling attention to the fence.

BALANCE

Another way to think of the distinction between formal and informal is summed up by the word "balance," or the idea that the fence style should match the house style. The critical importance of a good visual fit between the fence and its surroundings stems from the simple fact that, within the confined setting of a residential lot, a fence is inevitably the most commanding feature of the foreground or background. Imagine a suburban split-level ranch house with a traditional white picket fence: The juxtaposition doesn't make a lot of sense. And yet, the image of a post-and-rail fence seems to fit comfortably with either a traditional Cape or a ranch-style house.

Although picket-type fences are well matched to a wide range of historical styles, it doesn't mean that any kind of picket fence is appropriate for any style of old house. Try to observe historical accuracy. For example, the fanciful embellishment of the high Victorian style would appear as incongruous as a modern vertical-board fence when viewed in front of a Federal or Greek Revival home. A simple Greek Revival farmhouse may look garishly over-dressed with a very formal picket or board fence, while its elegant city cousin would require nothing less.

The principle of balance also applies to the inherent conflict between variety and unity. In general, the lines of the fence should act in concert with those of the house. But the overt echoing of the details of the house facade in the ornamentation of the fence could be very tacky. Sometimes, the overall impression is more pleasing when the dominant lines of the fence complement those of the house, as, for example, vertical pickets against horizontal clapboards.

CONTRAST

Understanding how to use contrast and rhythm is another requirement of this balancing act. Buildings and plantings will tend to stand out in contrast to a uniform fence. Design elements such as shrubbery and other plantings or lawn ornaments contrast with the empty spaces in which they're situated. These elements appear best when surrounded by an appropriate area of space. Because a fence divides as well as defines space, design elements that had seemed fine before the fence was built can suddenly appear crowded together when their capsules of open space are dissolved by the unifying visual field of the fence. Backgrounds and fore-grounds can become muddied, and the

Dark landscape features such as these trees and bushes contrast strongly with the crisp whiteness of the elegant picket fence.

A simple rustic fence complements the rough texture of the brick facade of the house behind it.

This fence merges with the stair railings to connect the house and street visually and architecturally.

focal point of the design can be lost. Large and small, dark and light, rough and smooth, uniform and variegated, the trick is to create interest, not irritation; harmony, not monotony.

The materials you choose for your fence should also be similar to those used throughout the neighborhood. Pay heed to regional fashions as well. If the majority of the fences in your part of the country feature unfinished or stained rough boards, it would take a talented designer to come up with a painted smooth-board fence that didn't seem out of place. Remember that a fence is by its very nature impossible to ignore.

CHOOSING A FENCE STYLE

As with most of modern building, the problem of deciding what kind of fence to build comes down to a surfeit of choices and a paucity of tradition on which to make them. We can pick from styles that run the gamut of historical epochs and span a world of cultures, but the result is often something akin to mixing plaids and stripes.

Short of hiring a professional designer, one way to avoid this kind of stylistic peccadillo is to go on a fence safari. Become a fencing field researcher. Armed with a camera or sketchpad, note not only fence designs that appeal to you, but also look at how they fit into the house and grounds. Try to analyze the gestalt of a pleasing fence: its height, width, and spacing of the pickets or rails, the scale of the

corner and line posts relative to the pickets and their detailing or lack of it. Does the infill consist of discrete panels that comprise regular modules, or does it extend in a continuous curtain? Does it recapitulate details of the house facade or contrast with them?

Look for other strongly repetitive themes and elements that fall under the designer's rubric of "rhythm" and "symmetry." For example, are the posts spaced so that the infill modules are evenly sized or does the fence end in an odd-width panel? If the panels aren't all the same size, are the different modules arranged in a regular order or otherwise

The repetition of posts and finials is the dominant rhythmic element of this fence.

This simple fence has an interesting rhythm. The short posts break the infill sections into regular bars, while the white pickets maintain a strong beat.

the case with simple rail fences. Finally, pay attention to how the fence makes its way across the terrain. If the ground is sloped, do the panels climb or descend in regular steps or do they trace the contour of the grade?

symmetrically balanced? How are changes in height handled? Is the transition gradual or abrupt?

Notice the gate (or gates). Does it seem to draw attention to itself or disappear modestly into the infill? Is it embellished with a decorative trellis? Many people find circular arches particularly evocative. Horizontal gateways suggestive of a Japanese *torii* are also highly resonant. You may discover that what excites you most is not the fence itself, but rather the plantings that it supports. This is often

PROPORTION AND SCALE

There are reasons why certain fences seem more "right" than others. I've just explored the concept of balance and hinted at some of its friends and relations. One of the most important of these is proportion. Something you may have noticed about the fences that struck you as more pleasing than others is that they seemed neither too tall nor too short. The height of their infill panels probably seemed suitable to their length. Is there indeed some ideal proportion between the height and length of an infill panel? Is there some reason why one size of rectangle seems more balanced than another? Of all the possible proportions between height and length, rectangles generated by the Golden Section of ancient Greece have traditionally been esteemed as the paragon of proportionality (see the sidebar on the facing page).

Another design concept closely related to proportion is scale, which can be understood as a rule of proportion for the relationships of parts to the whole. In practical terms, all this means is that you shouldn't mix big fences with little houses, or vice versa. Since the apparent size of an object

changes with the distance between it and the viewer, scale is not an absolute. The designer needs to consider the scale of a fence as a function of its location relative to its primary viewpoint. A fence that is too high when close to the house can seem just right at a distance. Another way of looking at this principle is that an object placed between a viewer and a more distant object can appear to increase or decrease the distance (see the photos on the facing page).

DESIGN CONSTRAINTS

In an ideal world, you'd be able to design your fence with only aesthetic considerations in mind. But in the real world, there are objectionable objects to screen out, unforgiving terrain and immovable objects to work around, and local fence laws to abide by. These and other constraints can have a significant effect on the design of your fence.

SIGHTLINES

The effectiveness of a privacy fence depends not only on the spacing of its infill and its height, but also on the distance between the observer and the view or object that you wish to screen out, or to put it another way, on the angle of your line of sight. Think of yourself, standing upright, as a vertical line that defines the angle of your sightline: The taller the object to be screened, the larger the angle of the sightline, and therefore the taller the fence must be to block it out. Distance between the observer and the object also affects the sightline angle since, due to perspective, tall objects appear to lower as they recede from the viewer.

To design a fence tall enough to screen objectionable views or objects from your point of view or to screen you from the view of outsiders, begin by studying your site, measuring off the necessary distances. Then, on graph paper, draw a horizontal line (assuming

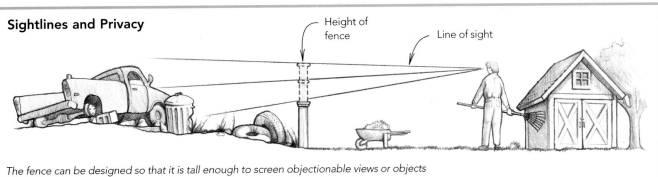

Sightlines and Privacy

Height of fence

Line of sight

The fence can be designed so that it is tall enough to screen objectionable views or objects from your point of view or tall enough to screen you from the view of outsiders.

your lot is flat, otherwise draw the actual contour). Plot the offending object and your desired point of view. The line that connects these dots represents your true line of sight. Now extend a vertical line up from the ground line at your proposed fence line until it intersects the sightline. Scale off the length of the vertical to find out how tall your fence has to be in order to block your view.

You can easily verify the prediction in the field. Stand or sit where you normally would. Have a helper stand on the proposed fence line with a plumbed story pole (a length of 1x2 will suffice) and mark the height at which the objectionable view disappears. Or, if working solo, plumb the story pole on the fence line, trim it

off at the intended height, and then sight across it from your viewpoint. If it appears higher than the object you wish to screen, the fence will be effective.

In Chapter 2 I discussed how fences can be used to screen out natural light for good or ill effect. But sunlight isn't the only kind of light that your fence might need to block. The light from artificial sources such as traffic, street lights, parking lots, a schoolyard, or nearby commercial and industrial areas can be very annoying. Study your site at night and plot the offending light source and your preferred point of view. Bear in mind that unless it's at a fair remove from your viewpoint, when the offending light comes from an elevated source such as a playing field or parking lot, it's unlikely that any fence

This low fence presents a rather unique solution to running a level fence across uneven ground.

of legal (or even practical) height will block it. Fortunately, designing a fence to block the relatively low-angle light of oncoming traffic is not normally particularly challenging.

A fence intended to enhance your point of view can sometimes cause problems from your neighbors'. A fence that may look good on your side of the line might plunge their narrow yard into a cold and gloomy shade. You might have to balance your desire for privacy with your neighbors' need for light and space. Ideally, a compromise can be reached by consulting your neighbor before you begin building. You might, for example, consider a short section of solid infill that gives maximum privacy in the most critical area and continues on as a trellis or a more open board or rail-type fence.

TERRAIN

The terrain of your site is a major design consideration. Other things being equal, building a fence on a flat and level lot is likely to be a lot less troublesome than building a fence across a slope or uneven ground. Slopes present both design and construction problems.

There are four basic ways to step a fence on a slope. The choice depends on the steepness of the grade and the kind of infill. The simplest option is to let the fence follow the contour of the terrain (see the top drawing at right). The posts are set plumb and equidistant but the stringers (or rails) are set parallel with the ground rather than

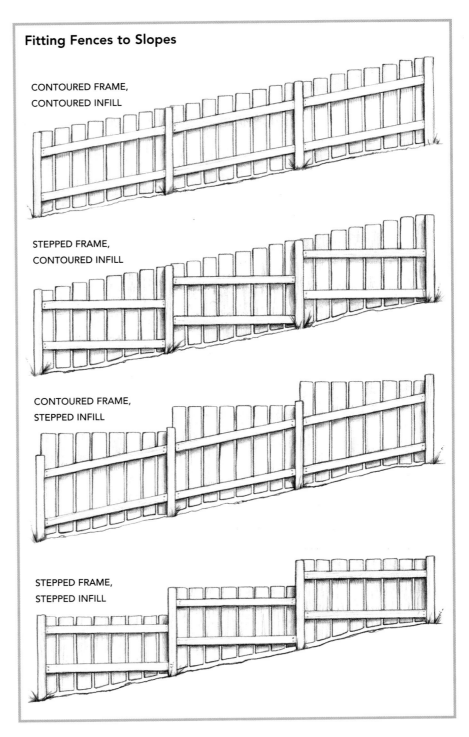

Fitting Fences to Slopes

CONTOURED FRAME, CONTOURED INFILL

STEPPED FRAME, CONTOURED INFILL

CONTOURED FRAME, STEPPED INFILL

STEPPED FRAME, STEPPED INFILL

The simplest way to fit a fence onto sloping terrain is to let it follow the contour of the land.

level. The infill is trimmed to project more or less evenly beyond the stringers at both the top and bottom of the fence—more on slopes that are relatively even and less on irregular grades. Depending on the kind of infill material, the effect of a contour fence is somewhat informal and, with grapestake or palings, even rustic. Contour fencing works equally well with post-and-rail or post-and-board fences and all types of vertical infill such as boards, pickets, palings, slats, and welded wire.

Contour fencing doesn't work for infills such as basketweave, horizontal closed boards, louvers, and inset panels. These types of fences, and any fence with sheet-type infill such as plywood or lattice, call for a stepped frame and infill design. Here the stringers are set level between the posts. The infill can be inset or else nailed on and trimmed to an even overhang at both top and bottom stringers. Either way, the rectilinear bays negotiate the slope in equal steps, maintaining a sharp distinction between the ground and the fence. Because of this formal and classic architectural quality, stepped fences are preferentially associated with elegant or historic residential properties.

The length of the steps (that is, the spacing between the posts) depends on the rise of the slope. If the step is too short, you'll use up more material than you need to. If it's too long, you'll be left with large triangular gaps under the fence. Not only are these gaps awkward-

Fences with lattice or solid infill fit sloping terrain best with a stepped frame and infill design. (Photo at left by Bernard Levine.)

One way to solve the visual problem of a sloping fence is to top the bays with concave (drape) or convex (arch) arcs instead of level steps.

looking but they also nullify any intended security and privacy functions. (This is also why a stepped post-and-rail or post-and-board fence tends to look unbalanced on any but a very gentle slope.) The steeper the slope, the shorter the step must be to keep the gap within acceptable limits.

Another way to overcome the problem of short steps on very steep slopes is to combine a stepped frame with sloping or contoured infill. The infill masks the large gap beneath the bottom stringer and transforms what would otherwise have been a series of jarring leaps at the top into a pleasing, gradually rising or undulating line.

I find that the eye can tolerate quite a bit of variation in the extension of the infill at the bottom of the fence as long as the projection at the top is relatively uniform.

OBSTRUCTIONS IN THE FENCE LINE

While there are a number of ways to accommodate a fence to a slope, other physical features such as immovable boulders and ledge outcroppings are non-negotiable site features. Your design options are simple: relocate or rearrange. If you have some flexibility regarding the exact location of the fence or the postholes, shifting the fence line

to sidestep the obstruction may be the easiest solution. This might also be a reasonable concession if you encounter more than a few boulders too big to dig up where your postholes are supposed to go (see pp. 151-153). When even post spacing isn't critical, you could simply shift the posts along the line and shorten or lengthen the stringers or rails accordingly.

What might at first seem like a problem could be an opportunity to add visual interest to your design. For example, "bumping" the fence out around a boulder or tree incorporates the erstwhile obstacle into the fence as a deliberate design element. As a bonus, the right-angle sections of this mini-garden or private nook act like pilasters in a long wall to resist wind pressure and sideways movement. In fact, "bump-outs" (or "bump-ins," depending on your point of view) are an elegant way to brace a tall closed-infill fence against high winds in any case.

The other option is to run your fence over the obstruction by cutting the bottoms of the pickets or boards to follow its contour. Typically, the bottom stringer must be raised to clear the boulder and maintain the minimum overhang of the infill at the highest point and a clearance of at least 2 in. between the bottom edges of the boards and the rock. These requirements can be troublesome if the rock happens to stick up high out of the ground. To forestall the increased likelihood of

Cutting a notch in the top edge of a fence to allow room for an overhanging branch is one way to deal with an obstruction.

warping, fence boards and pickets shouldn't project much more than 1 ft. beyond the stringer. Otherwise, it's a good idea to add a supplemental diagonal stringer to shorten the un-supported overhang. An arched stringer could be a graceful alternative.

Trees in the vicinity of your fence line can also present a problem. For example, it's generally illegal to cut a boundary-line tree. And even when

If you decide to make a tree part of your fence rather than build around it, run the fence up to the tree but not onto it. Never attach any part of a fence to a tree. Bacteria and fungi invading the tree through the wounds left by staples, nails, or screws driven into its trunk can cause the tree to sicken and die. Similarly, wrapping wire around the trunk eventually chokes off the sap flow and strangles the tree. Furthermore, setting fence posts too close to a tree can injure its roots, with the same unfortunate consequences.

To incorporate a tree into your fence, first protect the roots by setting the posts several feet back from either side of the trunk. Next, extend the stringers to within a few inches of the trunk. Then cut the infill boards to parallel the trunk and add a suitable end piece or a false post for extra support or a finished appearance as needed. Remember to leave enough clearance so that the tree doesn't grow into the fence. Otherwise, design an infill that can be easily pruned back as the gap shrinks.

The tree that had originally been considered a nuisance to the fence builder actually added visual interest to the design once it was decided that the fence would be built around it.

legality may not be an issue, it's hard to justify the felling of a mature and beautiful tree to make room for a fence. If relocating or bumping out the fence line isn't an option, consider making the tree part of the fence instead (see the sidebar on the facing page).

WEATHER

One other physical characteristic of your site that is important to consider when you design your fence (and one that's often overlooked) is the local weather. Intense sunlight causes boards to warp, split, and age prematurely. Paint or varnish peels, blisters, and fades. In damp or humid climates, water held against the fence invites the invasion of wood-destroying fungi and insects and disfiguring mildew. The grade and species of your lumber, its moisture content, the way the boards are attached to the frame—in short, your choice of materials, fasteners, finishes, and construction details—are all to a greater or lesser degree determined by their ability to withstand the ravages of the local climate. (For advice on choosing materials, see Chapter 4.)

The weather can also influence the structural design of your fence: When heavy rains turn hard ground into mud, fence posts that are set too shallow can shift out of plumb. Unless you add

Paint or stain not only makes a fence visually appealing but also helps protect it against the ravages of weather. (Photo by Jane Waterman.)

reinforcement or use a design that lets the wind and snow flow through, the pressure of strong and steady winds or drifting snow will inexorably push your fence over.

Those of us who live where the ground freezes in winter have to deal with the phenomenon of frost heave, which is what happens when water held in the soil freezes. Moving in the only unobstructed direction, the frozen soil expands upward, taking anything buried in it along for the ride. Repeated cycles of heaving and settling loosen the fence posts, and the fence begins to lean and becomes ever more vulnerable to

wind and snow. It would seem to follow, then, that fence posts should always be sunk below the frost line, but digging down that deep can be an insurmountable chore (see Chapter 6).

A humid or rainy climate exacerbates the problem of rot. Over time, water drawn into the bottom of the fence will move farther up into the wood. Besides underwriting the colonization of the shady and bottom-most sections of the fence by mosses and mildews, damp wood that cannot dry out soon begins to rot. Protecting your fence against rot might be a compelling reason to use pressure-treated lumber (see pp. 98-101).

Surface-water conditions on your site can also affect fence design. The most obvious and probably least common example of this principle is when a fence must span a stream, drainage ditch, or seasonal watercourse. For a pond or a wide and deep stream with a dependable flow, it usually works well enough simply to end the fence at or just beyond the water's edge. But for seasonal streams or gullies and washes subject to flooding and wildly fluctuating water levels, the fence has to maintain an effective barrier at low or no water and manage not to snag floating debris or be swept away at flood stage. The particulars of flood-fence design are obviously site-specific. Yet all flood fences feature one or more movable panels, hinged either at the top so as to float upward and outward, or at the sides to float freely against the banks as the water rises.

One of the ways a good fence design strives to minimize vulnerability to rot is by maintaining adequate clearance from the ground.

Whether the effect is intentional or accidental, the beauty of this fence is certainly doubled by its reflection in the pond.

EASEMENTS AND UTILITIES

In rural areas, a marginal portion of the land bordering the public roadway (the "berm") is actually under municipal jurisdiction. This easement gives the town a legal right to mow the roadside weeds and to knock over your mailbox with its snowplows. Just how far this authority extends from the center of the roadway varies according to locale. If you intend to erect a fence along the roadway, check with your town's highway department or road commissioner's office to find out how far back from the road to put it. In snow country, common sense alone would dictate setting the fence far enough back so that it won't interfere with plowing operations or be pushed over by the weight of the snowbank.

In more densely settled areas, the berm area is usually reserved for the

In some areas, you may have to set the fence back from the roadway.

sidewalk, and in some towns you may have to maintain an open strip of lawn between the walk and the street. Front-yard fences are only allowed on your side of the walkway. Check with the building or zoning department for any required minimum setback from the edge of the walkway or street.

Underground utility lines that run from the street to your house may belong to you outright, or they may be owned by the relevant public or private utility by right of easement. It's also not uncommon for utilities to cross your property to an abutting property. It may even be that a buried cable or natural-gas pipeline happens to run across your land. Any such easements are (theoretically) always noted in your deed and title-search documents and are also typically shown on maps located at the municipal offices along with the book of applicable regulations. You may need to obtain written permission from the utility to dig near an easement. In any case, heed the advice of any markers on your property that warn of buried cables before digging or drilling your postholes (see p. 150). Cutting through a buried power, water, telephone, gas, or sewer line can be a costly mistake.

Another type of underground utility that could interfere with your fence building, particularly in the Southwest, is a sprinkler or irrigation system. Besides having to move a supply pipe or relocate a distribution fixture, you should consider the effect of the existing spray pattern on the new fence. Water stains may detract from the appearance of the fence. Unless protected with treated wood or well-kept paint, the increased exposure to moisture could also shorten the life of the fence.

In addition to utility and municipal easements, the enjoyment and use of your land may be subject to other easements and rights-of-way that have to be considered in your fencing plans. A deeded right-of-way to an adjoining property is the most obvious case. Any fence you build would have to include a gate where it crosses the right-of-way. Since so-called "easements of necessity" cannot be blocked by law, you should discuss your plans with your neighbor first, secure his or her permission for the gate, and draw up a written agreement governing its operation and maintenance.

SUBURBAN FENCE LAWS AND COMMON COURTESY

Exactly where you place a fence on the boundary of your property may be governed by local laws. After you've decided to build a fence and before you make any specific design decisions or commit them to paper, contact your municipal planning office or building inspector to find out what requirements and restrictions apply to your fence plans and the appropriate procedure for obtaining the required permit.

The codes and ordinances governing fence construction vary greatly from community to town to state. All of them define basic design and construction requirements in order to ensure that whatever you propose to build doesn't unduly infringe upon the rights of others and the public good. Building codes typically define basic design and construction requirements, particularly from the standpoint of ensuring health and safety, whereas ordinances deal more with fostering conformity to community standards of appearance and safety.

Just about every urban and suburban municipality at the very least regulates the height and location of residential fences, including hedges and trees. Typically, backyard and side-yard fences are limited to 6 ft. or 8 ft. and front-yard fences to 3½ ft. or 4 ft. Most towns also stipulate setback

Some communities allow the maximum height of a fence to be exceeded if the fence panels are topped with lattice work.

requirements from buildings and property lines (except where the fence is jointly owned). Because fences can obstruct a motorist's view of oncoming traffic at a street intersection or create a blind drive or other hazard to the general public, special conditions normally apply to fences built on corner lots.

Some towns also prohibit the use of "hazardous" materials such as barbed wire. Others attempt to maintain minimum aesthetic standards and enforce upkeep through the power of blighted property ordinances to remove "eyesores." Most towns also require a

building permit (and associated fee) for any type of outdoor construction.

Even if your proposed fence will be entirely within your property, before you finalize your fence design and certainly before you dig the first posthole, common sense and neighborly etiquette suggest that you inform your neighbors of your intentions. Your neighbors could interpret your fencing activity as a hostile act, especially if you are a newcomer or fences aren't common in the neighborhood. They may (rightly or wrongly) interpret your intention to fence your children or animals in as fencing your neighbors out.

A fence that borders a street intersection should be kept low so that it does not interfere with the sightline of motorists entering the intersection.

The history of fence disputes is probably as old as civilization. The court records of the earliest Colonial settlements are full of suits brought by aggrieved planters against stock farmers over the depredations of their free-ranging livestock, and countersuits by stockmen over the shoddy and unsound condition of the planters' fences, which should have otherwise withstood their animals' investigations. The dockets also show frequent pleas for redress for injuries suffered when an argument between neighbors over the exact definition of a "sound" fence escalated into fisticuffs.

The potential for strife among the community prompted the selectmen of New England towns to institute the office of hayward or, as it was more commonly known, fence viewer, as early as 1643. These appointed officials, usually sporting a frock coat indicative of their high municipal office, were charged with walking the settlement fence lines to "see that

the fence be sett in good repaire, or else complain of it." They noted breaches and other defects and then informed the appropriate shareholders, who were required to make satisfactory repairs within the specified time limits or be subject to fines or other penalties.

The fence viewer was also the arbiter of whether or not a fence was sound. If the fence was declared sound, its owner could collect damages when an animal broke through it, but if judged unsound, the owner had no claim to "satisfaccon."

Although the office of fence viewer has gone the way of the frock coat, fence laws are anything but obsolete. There are still laws on the books in all 50 states intended to protect livestock from people and property, and vice versa. In some states, animal owners are required to enclose their animals or else be responsible for any damage they might do. In other states, animals are allowed to wander and fences are intended to keep them out.

In rural areas, neighborly relations still tend to be more important than legal rights. Nevertheless, informal understandings between neighbors won't hold up in court if challenged by a new landowner. You can do your own legal research by looking up "fences" in the index to the state statutes at the law library of your county courthouse or at city hall or the public library.

But, generally, if you don't keep animals, any fence you might build is unlikely to cause problems with the neighbors, so long as it's wholly on your own land and well within your property line. This is especially true if you follow the old country ritual of "walking the line." Viewing ambiguous boundary fences with your neighbor, particularly if you are the new neighbor, is a time-tested preventative against disputes arising over ambiguous boundary fences and any associated fence-building, land-clearing, or tree-cutting activities on the part of abutting landowners.

A little diplomacy can prevent a lot of ruffled feelings. Tell your neighbors what the fence will look like and ask for their suggestions. Stress how your well-crafted fence will enhance both the beauty and property values of the neighborhood. Try to be sensitive to local custom. A good design is one that harmonizes, not only with your own property but also with the rest of the neighborhood.

If all your efforts to enlist your neighbors' cooperation fail, you might decide that you really do want to fence them out. And, just for spite, you'll make sure that your fence sends the appropriate message. But be aware that some states have specific laws against "spite" fences; in many states, any fence over the proscribed height is considered a spite fence. Under common law, the fence builder's motive must be found to be "malicious," that is, the fence must serve no reasonable purpose to the owner other than to annoy or harm his or her neighbors or interfere with their use and enjoyment of their property. Even in states where no specific spite-fence laws are on the books, the court could still issue a halt-work order or force you to tear down the fence and compensate the plaintiffs for loss of use and enjoyment of their property.

FARM AND GARDEN FENCES

If you're planning to build a farm or garden fence, one final design constraint is exactly what it is that you're trying to fence in. Whatever the material used for the fence, a knowledge of the behavior of the animals it will contain is critical to understanding the best way to build it. As the old prescription "horse high and hog tight" for a legal fence suggests, different kinds of livestock react to fences in characteristically different ways. From the farmer's point of view, this meant that he had to build a different kind of fence for each kind of stock he kept on the farm. Nineteenth-century farmers typically employed a "trinity" of fence forms: a formal picket fence for the house and garden; board fences for barnyards and corrals; and rail or worm fences for the outlying pastures (see the print on p. 88).

Cattle fences Cows push on fences. They keep on pushing at the same spot until it weakens, and then they push through the fence or trample it down.

A fence that relied on strength alone to withstand this patient bovine battery was all too often outmatched by the animals' brute strength. A wood fence strong enough to hold in cattle needs massive, firmly anchored posts and heavy planked rails at least 4½ ft. high. If nailed or bolted to the fence, the rails should be on the inside to keep the animals from pushing them off. While the intelligence of cows is arguable, they at least know enough not to lean on something sharp. This is why a few strands of barbed wire strung between flimsy posts will restrain an 1,800-lb. steer more effectively than the heaviest rail or board fence.

Horse fences Horses are different. Since they invariably prefer the greener grass on the other side of the fence, no matter what color it may be on their own, they eventually break down wire fences as they reach over to get at the inviting delicacy. Barbed wire is as off-putting to horses as it is to cows, but horses have a hard time seeing it and will often maim or mortally wound themselves when they bolt full tilt into it.

For this reason, in the modern farmstead of wire fencing, a horse fence is still made of wide board rails, traditionally painted white for visibility. The posts and rails need not be quite as rugged as a cattle fence since horses don't lean on them. But they should be higher, at least 5½ ft., or higher than the horse's shoulder to thwart the over-the-top-and-under-the-line equine

This strong cattle fence ties 6x6 posts spaced 8 ft. apart with 2x6 plank rails.

Horse fences are constructed of wide board rails, traditionally painted white to make the fence visible to the animals it encloses.

Nineteenth-century farmers used different kinds of fences depending on the animal or area to be enclosed.

grazing habit. The specific number of rails and their spacing isn't important as long as they are "hoof-proof"—the spaces between the rails in the lower 3 ft. of the fence and between the bottommost rail and the ground must be big enough so that a pawing horse can withdraw its hoof without getting hung up (or else the spaces must be too small for the hoof to fit between). The top rail is often capped to make it stiffer.

Sheep and goat fences The problem with sheep fencing is not so much keeping the sheep in as keeping their predators out. The ability of welded-wire mesh to deter dogs and coyotes makes it the preferred fencing material among sheep people (barbed wire should not be used because it can tear a sheep's fleece). A rail or board fence would also work fine, as long as the bottommost rail is close enough to the ground to prevent a lamb from squeezing under it.

However, the acrobatic abilities of their predators requires an overall height for sheep fencing of about 4½ ft., and their persistence justifies a strand of barbed wire running over the tops of the posts.

Both welded-wire and board-rail fences are fine for goats, which aren't as vulnerable or as intimidated by dogs and coyotes, but the top of the fence should be at least 5 ft. high since goats can jump.

Hog fences When it comes to fencing, hogs are the epitome of barnyard incorrigibles. They can lift amazing weights with their snouts. Levered under a bottom rail, they've been known to topple entire fence panels. A determined hog can root under any fence or even squeeze between loose barbed wires. Hog fences should be stout, be between 3 ft. and 4 ft. high, and include

a strand of electric wire between the ground and the bottom rail.

Chicken fences Poultry fencing is like sheep fencing, only taller and with smaller holes. Six-foot poultry netting will keep the birds in and the vermin out. The bottom wire should be stapled to pressure-treated 1xs set in and staked to the ground to close the gap. Although it's done all the time, if the bottom of the poultry netting is buried and staked into the ground, it will soon rust. Run a strand of electric wire across the outside of the fence just above ground level to keep dogs and other pests from trying to dig under the wire.

Garden fences As with sheep fencing, the problem with garden fences isn't keeping the vegetables in but keeping uninvited connoisseurs out. A garden

Fences are the perfect backdrop for all kinds of gardens, from formal flower designs to wildflowers and crawling rose bushes.

A trellis gate can have a variety of functions. This one serves as an opening into the garden, a place to grow vine plants, and a unique design factor that adds visual interest for passersby.

fence is not engineered for brute strength; gardens are oases of delight, to the spirit and the palette, and their fences should please the eye first and foremost and then provide security. While looks are important, a garden fence should not block sunlight where needed and should provide shade where desired.

Closely spaced rough pales, grape-stakes, and split or whole saplings are all traditional "rustic" garden fence materials of variously effective deterrent power. The airier welded-wire fabric now most often used in their stead is long-lasting, easy to install, and, with attention to detail, can be good-looking and of similarly mixed containment capability. A 3-ft.- or 4-ft.-tall fence built from any of these materials will successfully repel nonclimbing pests like rabbits and skunks and usually all but a burrowing dog. But they all fail to keep out climbers and leapers such as the infinitely resourceful raccoon and the aerially acrobatic deer.

As for foiling the deer, the much-touted repellent effect of blood meal or the various other powders and pellets is either short lasting or worthless. The

The author's "maximum-security" garden fence keeps out all but the most determined animal intruders and also incorporates a decorative trellis for climbing vines.

only truly effective barrier against deer is an 8-ft.-tall fence. Poultry netting, or wire fabric, hung like a curtain from turnbuckled wire rope or stapled to board or peeled-pole rails makes a good, lightweight defense. Where I live, the local deer herd regards gardens as open-air cafés. No unprotected garden stands a chance against their marauding appetite.

My "maximum-security" garden is fenced with 8-ft.-tall 6x6 pressure-treated corner and gate posts, with 4½-ft.-tall 4x4 pressure-treated line posts on 10-ft. centers, supporting peeled, native-cedar top and bottom rails. A 4-ft.-high course of heavy-gauge welded-wire sheep fencing is stapled to the rails. A ³/₁₆-in. tensioned steel cable runs between the tops of the terminal posts, threaded through the top edge of a second 4-ft. course of light-gauge welded-wire fencing, which is likewise stapled to the upper rail at its bottom. Both welded-wire fences are electrically

connected and grounded. Single strands of hot wire at top and bottom discourage climbing raccoons and ground-level infiltrates.

Dog fences Even when dog enclosures are not required by law, owners of small lots and large dogs usually consider a dog fence an absolute necessity if they wish to maintain a habitable back yard. These enclosures are best built with wire mesh and pipe or standard chain-link fencing since most dogs eventually chew their way through boards and the average suburban homeowner has neither the affinity nor the need for electric or barbed-wire deterrents. The bottom of the fence should extend at least 1 ft. beneath the surface or inward on top of it (anchor it with tent pegs or rebar) to discourage the dog from digging under the fence. The gate should swing inward. This way there's less chance of the dog slipping past you and escaping than if it opened outward.

A single strand of thin electric wire is an effective deterrent against animals.

four

FENCE MATERIALS

A durable post is a matter of great importance. The red

cedar and the locust are the most enduring.

Split posts are always both stronger and more durable than

sawn ones. They are not cut across the grain, and the

water does not enter them so readily.

The Country Gentleman: A Journal for the Farm, the Garden, and the Fireplace (1862)

The unfortunate fact that wood in contact with moist earth soon rots has been the bane of fence builders ever since they first drove sharpened stakes into the ground. Historically, this problem has been addressed by the use of naturally rot-resistant wood species, by designs that kept the frame from direct contact with the ground or minimized surfaces on which water can sit, and by frequent applications of paint or other preservative coatings.

The introduction of pressure-treated wood has dramatically increased the useful service life of wood, but there are concerns about the health risks of treated lumber (as there are concerns about the environmental ethics of using increasingly scarce rot-resistant woods like redwood and cedar).

In this chapter, I'll examine the pros and cons of the various woods available to help you choose the right material for your fence. I'll also offer advice on

which fasteners to use to hold it all together and on how to choose between paints, stains, and clear coatings as a finish for your fence.

WOOD

Because wood is susceptible to decay and fence building is labor-intensive, it's hardly surprising that Colonial farmers would have had a strong incentive to discover and use the most

Whether painted or stained, untreated board infills will last longer if given some kind of protective coating.

rot-resistant species for their fences. What is remarkable, however, is that the early settlers, surrounded by species of trees unfamiliar or unknown in their native lands, cultivated their sophisticated working knowledge of the nature and properties of wood so quickly. In the span of hardly more than a generation, they had learned that the most durable species (in order of declining longevity) were locust, chestnut (or cypress along the southern seaboard), cedar, walnut, and white oak. Because oak was already familiar to farmers of English descent, this least resistant of the common post woods was preferentially used wherever it was found.

One advantage our ancestors had over us is that lumber cut from virgin forests was far superior to any available today. Old-growth lumber was strong, virtually knot-free, and, because the aromatic oils and resins in the heartwood repelled insects and fungi, highly rot-resistant. The ratio of heartwood to the looser, water-filled living cells of white sapwood in old-growth trees is almost opposite that of fast-growing trees harvested from third- and fourth-growth commercial forests. Today, few trees are allowed to grow old enough to accumulate much heartwood.

This explains why estimates of the life expectancy of fence posts made in the Colonial era are generally much higher than those for the same species today: For example, a report in a farm journal of the 1820s stated that white-

This rustic fence is made from spruce sawmill slabs and edgings. The heaviest rails are at the bottom, and the tapers alternate for a pleasing line. (Photo by George Nash.)

oak posts were still sound after 22 years in the ground, while current United States Department of Agriculture estimates give oak only a 7- to 15-year life. Decay resistance depends not only on the amount and quality of the heartwood, the diameter of the post, and whether it was split or sawn, but also on local and regional factors such as climate and the dampness of the soil.

Ultimately, there are simply so many variables that can affect the durability of wood that trying to pin down the "correct" number of years for a given species is futile. In my opinion, the tables and charts of decay resistance for common woods that you'll find in most fence-building books are essentially worthless. In the first place, no one would ever seriously consider using many of the species they list for fence

Unfinished eastern white cedar weathers to a handsome silver-gray.

posts or even fence boards. Second, the "decay resistance" for any given wood in general isn't the same thing as its life expectancy as a post in the ground, which is quite a lot less than its life as a board or fence rail. What's important, from the fence builder's point of view, is whether you can expect your fence post to last only a few years or at least a decade, and how well a wood will hold up when used for boards or rails.

HARDWOODS

As a general rule of thumb, all common hardwoods are unsuitable for fence building. They all have too little decay resistance, and most have too much dimensional instability (that is, they warp, split, swell, and shrink excessively) for outdoor use. Hardwoods are also difficult to nail and work. Although white oak (also known as "post oak") does have fairly good rot resistance, like every other commercially available modern hardwood, it's much too valuable (and much too costly) as finish lumber to split into fence posts. Even a stand of young oaks in need of thinning is worth more as firewood than as fence posts.

Note that my categorical proscription was against "common" hardwoods. Chestnut was once a common hardwood prized for its durability and ease of working and was so widely used for fence posts that it was already scarce even before the native trees succumbed to an imported blight at the turn of the century. Three other "uncommon" hardwoods—black

locust, osage orange, and red mulberry—have no equal for natural rot resistance (estimates range from 15 to 50 years) and, more important, no value as lumber. Unfortunately, the demand for fence posts made from these species so exceeded the sustainable harvest that they became virtually extinct by the late 1800s.

Nowadays, in certain parts of the country (most notably the Midwest) some of the original stands of osage orange and locust have rebounded sufficiently to support a small fence-post industry. You may be able to buy precut fence posts from your local farm-supply dealer at a price below pressure-treated wood. Be aware, however, that such posts are sold as they come from the tree, that is, round, usually with the bark on, and almost never sawn or squared. While natural posts are perfectly fine for use with barbed-wire or electric farm fences and rustic sapling-rail fences, they don't lend themselves to architectural or ornamental fences.

Because of its dimensional stability and rot resistance, teak, an imported tropical hardwood, might be a good choice for boarding an architectural fence where an elegant but natural appearance is a greater priority than cost.

SOFTWOODS

None of the common softwood lumber species (pines, firs, spruces, hemlocks, and larches) should ever be used for fence posts. However, with attention to jointing and fastening details and the regular application of a suitable protective coating, all these softwoods can be used for infill boarding, rails, and frame members. If kept dry, they can last a long time and are thus a durable and economical alternative to the more expensive redwood and western red cedar. Red pine and balsam fir are so vulnerable to decay that they should be avoided for any kind of fence work. Eucalyptus and tamarack, on the other hand, are quite durable, but their tendency to warp makes them unsuitable for boarding.

Redwood and cedar (as well as juniper and cypress) have a popular reputation for decay resistance and insect resistance, which has led many people to believe that they can be used not only for fence boards and rails but for posts as well. While it's true that the oils and resins they contain are naturally unpalatable to insects and fungi, it's important to remember that those oils are concentrated only in the heartwood. Their sapwood has no more resistance to rot or insect attack than any other softwood. Most of the redwood and western red cedar marketed today is cut from second-growth stands that yield relatively little heartwood.

Durability and cost notwithstanding, redwood and red cedar would still earn their well-deserved reputation as the queens of outdoor woods on beauty alone. Redwood is widely used because

Western red cedar makes a very attractive fence with a natural appearance, whether used in a conventional manner and topped by lattice (above) or as a sunburst (left). (Photos by Bernard Levine.)

it is very stable, easy to work, and has a deep, rich color that weathers to a handsome silver-gray. And red cedar, although lacking the natural insect resistance of redwood, also displays the same desirable physical attributes. Since its native range is much wider, red cedar tends to be somewhat less costly as well. In any case, both woods are affordable enough throughout northern California and the Pacific Northwest to be the natural choice for fencing and exterior siding in those regions. The durable all-heart grades are considerably more expensive than grades containing a mix of heartwood and sapwood.

Western juniper and its eastern cousin, eastern red cedar, are widely distributed, durable woods that have a long history of use in fence building. However, the growth habit of junipers is generally so convoluted and shrublike that the wood is only suited for posts to be used with barbed-wire fences. In contrast to its present-day appearance as a small and somewhat gnarled ornamental tree, eastern red cedar once grew straight and tall in virgin stands, but they were harvested to extinction for fence posts and rails.

Eastern white cedar, which is native throughout the Northeast and eastern Canada, is a common and affordable lumber species, comparable in cost to pine. Although even more rot-resistant than its red relatives, the wood is not at all bug-proof. But it does display the other family virtues of workability,

stability, and good looks. It is an excellent choice for unpainted fence boards and pickets (see the photo on p. 96). Small cedar logs (3 in. to 6 in. in diameter) are used as is for farm fence posts and, when peeled, for split, sawn, or peeled-pole fence rails.

Of the common softwood lumber species, eastern hemlock is worthy of special mention. Although the wood is rather splintery and tends to split when nailed near an edge, hemlock is usually the least expensive wood you can buy from your local sawmill. It's a bargain hunter's choice for strong, rough-board fences and, if planed and dry, will accept paint or stain as well as any other common softwood.

PRESSURE-TREATED WOOD

Wood decays because certain kinds of fungi and insects like to eat it. In order for your fence post to be the featured entree on the menu the table must be properly set: Sufficient moisture and oxygen, a suitable temperature, and an adequate food source must all be present.

Wood can't rot if its moisture content is less than 20%. But in contact with moist ground, wood continuously absorbs water until it reaches the fiber saturation point (a very inviting 30%). This is why almost any softwood is immune to rot in an arid climate. Likewise, the lack of oxygen explains why wood continuously immersed in water won't decay either (although it is

eaten by various marine borers) and why fence posts tend to rot first and worst just beneath the ground level.

Fungi are dormant at temperatures below 40°F and they die at temperatures above 110°F, so the rate and likelihood of decay varies with the seasons. But some termites build tunnels to bring vital water into their wood mines, and certain beetles feed preferentially on dry wood. The point is, moisture, temperature, and oxygen content are, if not actually impossible, at least impractical to control. Fortunately, the food source isn't. Treating wood with chemicals that are toxic to fungi and insects effectively takes it off the table.

There are two problems associated with the chemical treatment of wood: first, how to get enough preservative into the wood for long-lasting protection; and, second, how to poison the pests without harming other forms of plant and animal life, including the fence builder.

Surface coatings are just that—they go *onto* wood, not *into* it. And whatever limited protection a brushed-on coating does provide is soon compromised by the splits and cracks that open the unprotected interior of the wood to attack as it swells and shrinks with changes in moisture. The coatings also simply flake off and wear away. Recoating the buried portion of a post isn't a practical option. Although dipping or soaking the wood is about four times more effective than

The deep brown color of this specially treated wood needs no other finish for a handsome fence. (Photo by Huck DeVenzio.)

In the commercial process employed today, wood is pressure-treated in giant cylindrical steel pressure vessels that force the preservative chemicals deep into the wood cells. (Photo by Huck DeVenzio.)

Pressure-treated wood has been subjected to extensive performance and safety tests ever since chromated copper arsenate (CCA) was registered as a pesticide in 1947. In 1984, the Environmental Protection Agency (EPA) concluded an eight-year study of the merits and unacceptable risks involved with the use of this preservative. By law the EPA is also required to reevaluate all registered pesticides every five years.

The latest studies support the conclusion that CCA-treated wood is suitable for all recommended uses without restriction. (A toxicity study claims that a horse would have to consume the equivalent of a 10-ft. 2x4 in one sitting to receive a lethal dose of preservative.) Nevertheless, the EPA does recognize that exposure to wood preservatives may present certain hazards and so recommends the following precautions for handling, using, and disposing of treated wood:

• Avoid frequent or prolonged inhalation of sawdust.
• When sawing or machining, wear a respirator or dust mask.
• If possible, perform these operations outdoors to prevent indoor accumulation of sawdust.

• Wear safety glasses to protect your eyes from flying particles.
• After working with or handling pressure-treated wood, and before eating, drinking, or smoking, wash hands thoroughly.
• If sawdust accumulates on clothes, launder separately from other clothing before reuse.
• Dispose of treated wood in the appropriate bin at your local transfer station or include with your normal trash. Never burn in open fires, stoves, fireplaces, or residential boilers. Toxic chemicals may be produced in the smoke and ash.
• Do not use where the preservative can be in prolonged contact with food or animal feed (i.e., food or silage storage structures). It's only prudent to take precautions to minimize direct exposure and contact. I know from personal experience that a splinter of treated wood is much more irritating and quicker to fester than ordinary softwood. At the very least, I always wear work gloves if I have to handle any appreciable amount of treated lumber.

Wear a respirator and safety glasses when sawing pressure-treated wood.

brushing, the preservative seldom penetrates more than a tenth to an eighth of an inch.

Traditional oil-borne wood preservatives such as creosote oil and pentachlorophenol are extremely hazardous. In 1934, an English patent was issued for a preservative treatment using water-borne oxides of arsenic and copper. Subsequent improvements in the formulation and technology led to a process that produced virtually rot- and insect-proof wood that was also free of oily residues, odorless, nonstaining, safe and nonirritating to handle with minimal precautions, and not harmful to humans, domestic animals, wildlife, plants, and the environment.

Although the original formulation for pressure-treating wood used potassium or sodium dichromate and copper sulfate, oxide formulations were found to produce wood without the objectionable and potentially corrosive and electrically conductive surface residue that tempted animals to lick and chew it and was nasty to handle to boot.

In the process employed today, wood previously dried to 25% or less moisture content is loaded onto a dolly and wheeled into a huge cylindrical steel pressure vessel (see the bottom photo on p. 99). A vacuum pump exhausts the air from the cylinder and the wood cells as well. Then the tank is filled with preservative solution. A pressure pump forces more preservative deep into the wood cells until they can absorb no more. The excess solution is pumped back into the storage tank, and the wood is removed and then re-dried in kilns or left to season in the open air.

The most commonly used preservative is chromated copper arsenate (CCA). The copper delivers the antifungal effect (and imbues the wood with its characteristic pale-green tint), chromium gives it some extra punch and is critical to the fixation reaction, and the pentavalent arsenic oxide kills the bugs.

Not every species of wood is amenable to the treatment process. In some, the wood cells soak up the preservative like a sponge. Others are simply too dense to absorb any appreciable amount of solution. In general, almost all pines, most western fir and larch species, coastal Douglas fir, redwood (the sapwood not the heartwood), and oak have the required porous cell structure. The absorption capability of some denser woods (such as Douglas fir) can be improved by perforating the surface.

Because it combines great natural strength with a sponge-like cell structure capable of absorbing up to 4½ gal. of preservative solution per cubic foot, southern yellow pine (SYP), a fast-growing species harvested from plantation forests, is the hands-down favorite for CCA treatment. In fact, about 45% of all the SYP harvested each year ends up at the treatment plant, and roughly 12% of that goes to building wood fences.

Ultimately, the choice of any building product almost always entails some sort of trade-off between the benefits and potential risks associated with its use. Depending on your point of view, the choice to use pressure-treated lumber may be the lesser of two evils. Notwithstanding the skepticism with which many would greet the positive safety studies of the Environmental Protection Agency, whatever misgivings or outright objections one might have concerning the environmental or health risks of CCA-treated lumber must still be weighed against the fact that using treated wood does offer some undeniable environmental benefits, especially in light of the increasing scarcity of its "natural" alternatives. Because pressure-treated wood will resist decay for at least as long as it would take to grow the lumber needed to replace structures destroyed by rot or insects, fewer trees have to be cut. According to one estimate, using pressure-treated wood conserves 6.5 billion board feet (the equivalent of 425,000 new homes) of wood annually. Since about 167,200,000 board feet of lumber go into building wood fences each year, using pressure-treated wood can save a lot of untreated lumber for other purposes.

The preservative retention level of pressure-treated wood (as indicated on the treater's mark stamped or stapled onto the lumber) should correspond to the intended use. If the label on the wood reads "0.25" (lb./cu. ft.) or "LP-2," the wood is suitable only for "above ground" use. While that might do for fence rails or infill boards and pickets, wood used for fence posts should bear the 0.40 (or "LP-22") "ground contact" marker if you want it to last.

If available (not all suppliers carry the full range of products), treated wood rated at 0.30 (intended for above- and below-ground use for lumber less than 4 in. thick) is a more durable alternative to the lower-rated wood. Higher concentrations of preservatives (up to 2.50) are called for only when the wood will be immersed in water. One caveat: Beware of pressure-treated wood labeled "treated to refusal," that is, until it refused to absorb any more preservative, which if the wood was already full of water before treatment, wouldn't leave much room for preservative.

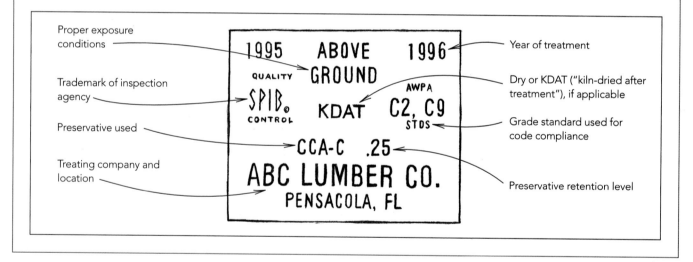

Proper exposure conditions

Trademark of inspection agency

Preservative used

Treating company and location

1995 ABOVE 1996
QUALITY GROUND
SPIB® AWPA
CONTROL KDAT C2, C9
STDS
CCA-C .25
ABC LUMBER CO.
PENSACOLA, FL

Year of treatment

Dry or KDAT ("kiln-dried after treatment"), if applicable

Grade standard used for code compliance

Preservative retention level

CHOOSING WOOD

Treated or not, the grade of wood you select depends on the intended application. Since softwood lumber can be graded according to appearance, appropriate use, structural properties, or a combination of these factors, the whole business of lumber grades can be pretty confusing even to professional carpenters.

Fortunately, as a fence builder, you needn't be concerned with arcane tables and charts. All you need to keep in mind is that *dimensional lumber* (two-by anything and up) is generally graded according to its structural strength, while *boards* are graded according to their looks (see the chart on p. 104). The relevant concept is simple: The fewer obvious defects a piece of lumber has, the more it costs. This translates into selecting the grade that makes the best-looking fence for the lowest cost. You don't want to spend any more than you have to and no less than you should. While your fence can't look too good, it can certainly look too poor. So choose the best lumber that makes sense for your project.

One defect that isn't considered a grading criteria but that should be considered if children are likely to climb on your fence is how splintery the wood is. The face and edge surfaces of some woods, particularly eastern hemlock, tend to throw off splinters more readily than others.

In addition to grade, there are at least two other important considerations that enter into your choice of fence wood. The first is moisture content. Normally, "green" lumber, fresh from the sawmill, is stacked in stickered piles (with narrow strips of wood between the boards) to dry out in the open air over a period of several months ("dry" or "D" on the grade stamp), or placed in kilns that achieve the same results in a few days with less lumber lost to warping (stamped "KD"). Wood that is full of water is not only heavy but also, depending on the species, will warp and shrink considerably as it dries. It also won't accept oil-based paint, stains, or preservative coatings very well.

Although it may seem like a reasonable cost-cutting measure, any savings gained by using green lumber in your fence project will be offset by the problems created when boards split apart around fasteners and cup and twist off their rails, when gaps appear between tightly butted joints, and when nails "pop" out of the surface of the wood.

Note that pressure treatment doesn't inhibit the natural tendency of wood to warp, split, cup, or twist. All wood is hygroscopic, that is, it absorbs moisture from the atmosphere until it reaches equilibrium with its surroundings. Consumers, and even worse, suppliers who ought to know better, assume that because pressure-treated wood resists decay, it can be left stacked in unstickered piles outdoors. This is a mistake. Water trapped between the boards soaks into the wood, negating any seasoning and kiln-drying, which is why treated lumber usually feels so heavy. It's also the reason why pressure-treated lumber can warp so severely once it's put into service. Keep this in mind when you select your wood: Just because the grade stamp says KD ("kiln-dried") or S-Dry ("surfaced dry") doesn't mean it stayed that way. The best way to avoid these problems is to buy wood marked KDAT ("kiln-dried after treatment") and sticker it so that it can't reabsorb moisture.

The second factor to consider when choosing a fence wood is the surface finish. One shouldn't confuse "green" lumber with "rough" lumber. The term *rough* refers to surface treatment, not moisture content. Rough lumber can be either dry (usually air-dried) or green. Seasoned rough lumber is considerably

This unusual fence is made from wide, waney-edged red cedar slabs ("culls") that would otherwise yield few square-edged boards. (Photo by Charles Miller.)

LUMBER GRADES

SAMPLE CHARACTERISTICS		NO. 1	NO. 2	NO. 3
		Maximum allowances for each grade		
	Knots			
	2x4	1½	2	2½
	2x6	2¼	2⅞	3¾
	2x8	2¾	3½	4½
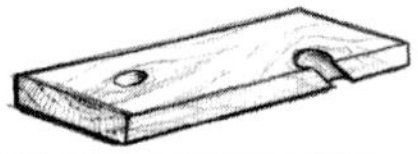	Holes	1 per 3 ft.	1 per 2 ft.	1 per 1 ft.
	Crook			
	2x6x10	5/16 in.	7/16 in.	5/8 in.
	2x8x16	9/16 in.	3/4 in.	1⅛ in.
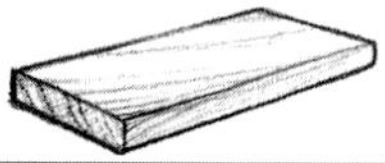	Slope of grain	1 in 10	1 in 8	1 in 4
	Wane			
	full length	¼ thick x ¼ width	⅓ thick x ⅓ width	½ thick x ½ width
	for ¼ length	½ thick x ⅓ width	⅔ thick x ½ width	⅞ thick x ¾ width
	Splits	width of piece	1½ width of piece	1/16 length of piece
	Shakes			
	surface	2 ft.	3 ft. (or ¼ length)	well scattered
	through	none	2 ft.	⅓ length

Adapted from Timber Products Inspection, Inc. (courtesy of Hickson Corporation).

less expensive than seasoned surfaced lumber (stamped "S," also known as "planed," "milled," or "dressed" lumber), especially when purchased at a local sawmill. There is one drawback to working with rough lumber, however. Rough is actually shorthand for rough-sawn, and, depending on the setup of the sawmill and the skill of the sawyer, there can be a considerable variation between the actual and nominal dimensions of the lumber, between individual pieces, and even from one side or end of a piece to the other. The differences are likely to be small with the lumber produced by the laser-guided saws of large commercial sawmills; but the worn bearings and wobbly circular sawblade at your local mill can turn out what some rural builders refer to as "four-dimensional" wood.

Keeping butted fence boards that taper arbitrarily in width or thickness from end to end from running off plumb requires constant adjustment and checking with your level. Variations in thickness preclude a crisp, finished look and make it difficult to fit neat joints. On the other hand, the impossibility of precision and "clean" lines can just as easily be an asset as a liability when a rustic appearance is exactly what you want (see the photo on p. 103).

Fortunately, there's even a way to have your rough-looking fence without the attendant hassles of working with rough wood. You can have the fence boards surfaced on three sides, thereby eliminating the dimensional problems while preserving the roughsawn face.

DECODING A LUMBER-GRADE STAMP

Understanding the symbols on a grade stamp will help you talk like a native on your forays to the lumberyard. Grade stamps indicate the moisture content of the lumber as well as the surfacing treatment it has been given. The standard markings are as follows:

- S-Grn: "Surfaced green," which is planed green lumber.
- S-Dry: "Surfaced dry." Air-dried ("seasoned") to 19% moisture content (M.C.), this is the most common and least expensive grade of "dry" lumber sold.
- KD: "Kiln-dried" to a M.C. ranging from 12% to 19%. Although it costs more, it may be worth it if you need stable lumber that won't warp, split, or cup.
- KDAT: "Kiln-dried after treatment." This category is reserved for pressure-treated wood and is the grade you should use for the same reasons you would select KD wood instead of green or S-dry.

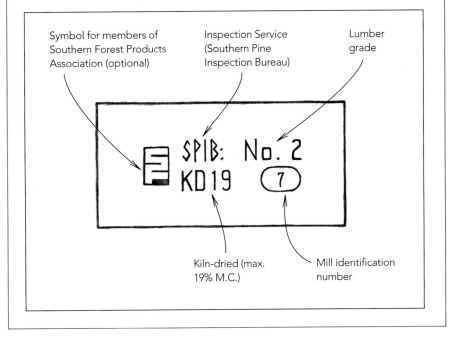

Symbol for members of Southern Forest Products Association (optional)

Inspection Service (Southern Pine Inspection Bureau)

Lumber grade

Kiln-dried (max. 19% M.C.)

Mill identification number

PREFABRICATED FENCES

Even though this is a book about building your own fences, you don't necessarily have to build them from scratch. Your local building center likely carries a large selection of prefabricated fence panels and fencing systems. And many fence contractors' showrooms also feature samples of premanufactured panels. At the very least, you could be inspired to try an unfamiliar pattern. And you could be pleasantly surprised to discover that installing a pre-manufactured fencing system is usually considerably less expensive than building your own fence from scratch.

Fencing contractors may also be your only source for specialty materials like split cedar rails. Although they would prefer to do the entire job, most fencing contractors would still rather profit from selling you the ready-cut materials for a do-it-yourself assembly than make no money at all. But, as always, remember that you usually get what you pay for. The least expensive prefab panels are often assembled from flimsy and poorly seasoned wood, with stapled fasteners instead of screws, and soon warp and twist and fall apart. And remember, also, that saving money isn't the best reason to build your own fence. The bottom line is that you can always buy a prefab fence for less than you can build a custom one. But a prefab fence won't be *your* fence.

WOOD-POLYMER LUMBER

At the time of this writing, a somewhat promising new product has been introduced into the fencing market that some feel may offer a palatable alternative to treated wood. "Trex" wood-polymer lumber is a 50/50 mix of recycled plastic (primarily grocery bags and shrink wrap) and hardwood waste molded into a material that is highly resistant to moisture, decay, insects, and ultraviolet damage. It can be sawed, sanded, nailed, and drilled like natural wood.

Trex wood-polymer lumber has a slightly roughened surface that will readily accept paint and stain, once the material has faded from its initial brown shade to light gray after 6 to 12 weeks of exposure to sunlight. It contains no toxic preservatives and can be disposed

Prefabricated or ready-cut treated-wood fence components can be assembled into an attractive fence. (Photo by Huck DeVenzio.)

of with normal construction debris. When burned, its smoke is no more toxic than ordinary untreated wood.

It would seem that such a material has all the benefits of pressure-treated wood without its potential dark side. The 1x6 and 5/4x6 boards and 2x2 balusters could be an excellent infill material. The product is also available in 4x4, 6x6, and 6x8 stock that could be suitable for posts.

The downside is that wood-polymer lumber isn't strong enough to be used for structural members such as joists, columns, or support posts. This rules out its use for stringers and rails, and for fence posts that will be subjected to strong lateral loads (such as a solid infill panel). The competitive edge of wood-polymer lumber is also somewhat dulled by its high cost—at this time, roughly twice that of pressure-treated wood (though about half that of premium redwood or red cedar).

There's also some evidence that boards can sag over time. Although it readily accepts screws and power nails, it's much harder to hammer nails into wood-polymer lumber than into ordinary wood, especially in cold weather when the material becomes brittle. And while it does crosscut easily, ripping down the length of a piece can be problematic, even with the recommended fine-tooth carbide blade. Finally, the verdict isn't in yet on longevity. The present 10-year warranty is probably conservative, but it certainly falls far short of the equally conservative 40-year warranty for treated wood.

ESTIMATING MATERIALS

Estimating the materials you'll need for a fence is a fairly straightforward procedure. To estimate how much wood to buy, you'll first need to a draw a "typical section," which is nothing more than a large-scale (¾-in. or 1-in. to 1 ft.) elevation of a single fence bay. Multiplying the number of individual components that comprise the typical section by the number of sections as shown on the site plan (see pp. 121-124) tells you how many of each component you'll need. It's a good idea to add about 10% to the total as an allowance for mistakes, unusable pieces, and other such "fudge" factors.

Of course, the bill of materials for a wood fence includes more than just wood. There's gravel and/or concrete for anchoring the posts, fasteners for assembling the fence, hardware for gates, and paint or other finishes—not to mention rental of power equipment or special tools, as well as sales tax.

PLASTIC LUMBER

Wood-polymer lumber shouldn't be confused with plastic wood. Panelized fencing, mainly of polyvinyl chloride (PVC, also known as "vinyl," the same stuff that gutters and plumbing pipe and siding are made from), is now being aggressively marketed as a "maintenance-free" alternative to old-fashioned wood fencing. Proponents also tout its splinter-free clean appearance and "durability." Since most of the vinyl fencing currently on the market offers only a 10-year warranty, the latter claim is a bit specious.

One major problem with vinyl fence panels is that they cannot be repaired. If a post or picket breaks, the entire section has to be replaced. No one claims that vinyl fencing is less expensive than wood fencing either. But, ultimately, the real problem with PVC fencing is the same as the problem with vinyl siding and gutters: There's simply no justifiable reason to turn nonrenewable petroleum into fencing or siding when perfectly adequate and renewable alternatives already exist. If you don't want to paint your fence, stain it or let it weather. If you are worried about replacing it, use pressure-treated wood, which costs a lot less and will probably last longer than you will.

FASTENERS

Without fasteners, all you've got is a pile of very nice fence boards. In addition to nails, screws, and staples, metal brackets, hangers, and anchors are frequently used in fence construction. It's important to use the right kind of fastener because, more so than with most outdoor construction, the extreme fluctuations in moisture and temperature to which a wooden fence is exposed severely stress the wood. As boards shrink, cup, and curl, the fasteners that secure them to the rails loosen and pull out.

NAILS

The ordinary common (or "bright") nail has no place in outdoor construction of any type, let alone in your wooden fence. In the first place, it's made from uncoated steel wire and will soon rust when exposed to the weather. Since rust is the metallic equivalent of wood rot, the damage is structural as well as cosmetic. Second, the common nail's smooth, thick shank, while resistant to bending stresses, has relatively little holding power.

Although the zinc coating of a galvanized steel nail does prevent rust, there are some limitations to its protection. Not all galvanized nails are equal. The thin coating of the cheap electrogalvanized (plated) nails often sold at discount builder's outlets easily flakes or wears off. A single hammer blow is enough to knock the coating off the nail head, leaving the most vulnerable part of the nail exposed to the elements. Electrogalvanized nails also have no more resistance to withdrawal than common nails. Look for "hot-dipped" galvanized nails instead. (Check the letters on the side of the nail carton: "E.G." signifies electrogalvanized; "H.D." stands for hot-dipped.) Their rough, heavy coating is long lasting and better able to withstand hammer blows. As any carpenter who has ever tried to pull one out will swear (often quite literally), the barb-like coating increases the holding power of the shank so much that the head will snap off before the nail budges. This suggests another even more effective way to increase the holding power of any nail: Use spiral-threaded (annular) or ring-shanked nails.

In my experience, the best nails for securing fence boards and pickets to stringers and rails are stainless-steel ring-shank siding nails. They hold so well that they'll pull clean through a picket or board before they'll pull out of the underlying rail. Since stainless steel is both rustproof and nonreactive, the nails don't cause stains (a problem with hot-dipped nails used in unpainted wood). Finally, the heads have a small, unobtrusive profile, similar to that of a wood-shingle face nail. Since nail heads "read" as markings on the page of the infill, larger head profiles can detract

Nails used for fence building should have high resistance to corrosion and withdrawal and a low tendency to split the wood. (Photo by Scott Phillips.)

from the overall appearance of the fence. The only real drawback of stainless-steel nails is their high cost, about three to four times greater than H.D. galvanized. But since 4 lb. or 5 lb. will easily suffice for more than 100 ft. of picket fence, the overall difference won't break your budget. The only other problem may be finding the right lengths or finding them at all at your local lumber dealer.

Many novice builders assume that when it comes to holding power, more nails are better than fewer nails. The counterintuitive truth is that beyond a certain optimal number, driving more nails actually weakens the strength of the connection. If you think of the tip of each nail as a tiny wedge that's splitting the wood fibers apart, the reason is obvious. Driving a series of

small closely spaced wedges is exactly the same technique stone cutters use to split granite blocks. Don't use any more nails than you need to. A good rule of thumb (for face nailing) is to use two nails for a 2x4, three for 2x6s and 2x8s, and four for 2x10s and 2x12s.

This same splitting action is also the reason why carpenters routinely blunt the tip of a nail that will be driven close to the end of a piece: A blunted nail crushes the wood fibers rather than splitting them.

No matter how many nails you use, they should be the correct length. Too short and your boards will pull away from their rails; too long and you'll split them or pierce clean through the back side. A good rule of thumb here is that the nail should be long enough to go completely through the top piece and at least one-third but not much more than half-way through the bottom piece. For example, for a 1-in. (nominal thickness) board nailed to a 2x4 on edge, use 4d nails. If the 2x4 backing is flatwise, use a 5d or 6d nail. Nailing a 2x4 rail against a 4x4 post calls for a 10d nail. However, if that same 2x4 were to be nailed to the top of the post, you'd use a 16d nail to counteract the poor holding power of nails driven into end grain. Since the diameter of a nail increases with its length, blunting the head of a large nail driven too close to the end of a piece might not prevent it from splitting. In cases like these, it's safer to predrill a slightly narrower hole first.

Except in those few instances where joints are subjected to severe shearing stresses, I always use box nails instead of common. Their thinner shanks minimize the chances of splitting, both for end-nailed joints, and more important, with toe-nailed (nails driven diagonally through the end of a piece into the face of the other) pieces.

SCREWS

When it comes to ultimate holding power, nails can't compete with screws. Ring-shank and even spiral-shank nails both tend to mash their way through the wood, whereas screws cut themselves a path. Ordinary wood screws are both costly and time-consuming to use, and their standard electroplated zinc or chrome coatings soon rust. However, coinciding with the explosive growth of the residential pressure-treated wood market, the bugle-head screws familiar to drywall installers began appearing decked out in hot-dipped zinc coatings for outdoor use (the so-called "decking" screw).

Compared with wood screws, the thin shank, coarse thread, and very sharp point of bugle-type screws allow them to be driven rapidly and easily without predrilling a pilot hole (except near to the end of a piece, or in very narrow strips, where they otherwise will split the wood). Decking screws are also available in stainless steel (useful when durability is more important than economy, e.g., where there is exposure to salt water or air), and with a golden-hued cadmium plating intended for use with pressure-treated decking.

Although they cost at least as much if not more per pound than stainless-steel nails (and there's a lot less of them in a pound, too), standard galvanized

Decking screws, which have hot-dipped zinc coatings for outdoor use, are ideal for fastening fence infill to the rails. (Photo by Scott Phillips.)

screws are preferable to nails for fastening thicker infills such as stockade palings, split poles, and 2x2 balusters or dowels to their rails. Since screws turn themselves into the wood, the bottom board won't "bounce" under the steady pressure of a screw gun as it does when you try to hammer a nail home. Since the head of a screw is similar in diameter to that of an 8d box nail, you should give some consideration to the visual impact of the use of screws.

When you're installing hardware, always use standard flat-head wood screws rather than bugle-type screws, properly sized so that their tapered heads fit the screw-hole countersink and the diameter of their shanks is barely less than that of the screw hole. If the screw hole lacks a tapered rim, use round-head wood screws instead. To ease driving and prevent damage to the screw head, always drill a pilot hole first. Likewise, for the same reasons, use Phillips-head or square-drive screws instead of slotted-head screws, even if you have to buy extra screws to replace the ones packaged with the hardware.

OTHER FASTENERS

Depending on the scale of your fence, you may also have recourse to heavy-duty fasteners such as carriage bolts, machine (hex) bolts, and lag screws (also known, interchangeably, as lag bolts). Machine bolts are easier to tighten than carriage bolts since their hex heads fit a crescent wrench or socket wrench. However, round-head carriage bolts present a more finished appearance. Since the friction between the squared-off underside of the head and the wood it's drawn down into is what enables you to tighten the nut, be careful not to overtighten a carriage bolt. Otherwise, the entire bolt will spin in its hole when you try to back off the nut.

To maximize holding power, always insert washers under the head of a lag screw or machine bolt (and also between the nut and the wood for all types of bolts and screws). Likewise, when drawing two pieces together with lag screws, drill the pilot hole in the first piece slightly larger than the bolt diameter. Otherwise, the screw tends to push the top piece upward as it is turned down into the lower piece. Finally, as with plated nails, for best

Heavy-duty fasteners include (from left) a lag screw, a hex-head machine bolt, and two carriage bolts. (Photo by Scott Phillips.)

results, use hot-dipped bolts and nuts (which have a pitted dull gray appearance) instead of ordinary (smooth and shiny-looking) plated hardware.

POST ANCHORS
AND FRAMING ANCHORS

Post anchors (also known as post bases) are set in fresh concrete or bolted to existing slabs or columns. The fence post is then screwed or bolted to the anchor. To protect against rot, use an anchor that isolates the bottom of the post from direct contact with the concrete. Since the post isn't in contact with the ground either, it's not as vulnerable to decay and so you can use untreated wood. And, if and when the post does rot, you won't have to dig it up to replace it with a new one.

There is one considerable drawback to this otherwise seemingly ideal system: A post that's buried at least a third of its length into the ground or concrete is firmly braced against lateral forces. But a post set on an anchor is only braced at several discrete points by screws or bolts along a rather short portion of its overall length and, therefore, is correspondingly more vulnerable to sideward pressures. Rather than being a firm foundation, the anchor is actually the structural equivalent of a rusty hinge. To increase the resistance of this connection, secure the post to its anchor with two continuous through-bolts rather than lag screws. The bolts should be heavy enough to resist bending themselves: Use $\frac{3}{8}$-in. bolts for 4x4 posts and $\frac{1}{2}$-in. bolts for 6x6s.

Post anchors are used at the base of posts to isolate the wood from the concrete. Various framing anchors can be used to make a strong connection between fence rails and posts. (Photo by Scott Phillips.)

Steel framing anchors and post caps make a much stronger connection between fence rails and posts than does toenailing. They're a lot faster and easier to install than cutting dadoed or mortised joints, but this convenience is gained at the expense of appearance. Unless the anchors will be hidden beneath the infill boards, they will seriously devalue the integrity of an otherwise crisp design. Finally, various "L" and "T" brackets can be used to stiffen the corners of a gate frame and help it remain square over the years.

FINISHES

To paint or not to paint? The answer encompasses an intricate a priori linkage of social custom and convention, personal taste and preference, environmental ethics, and practical concerns and consequences. By the time Tom Sawyer came to stand, brush in hand, before the indifferent expanse of the most famous fence in American literature, the whitewashed fence had long been established as an icon of the American cultural and physical landscape.

During the pre-Revolutionary era, fences throughout the Colonies were typically painted white or yellow to match the color of the house. Often, since real white and yellow paint (as opposed to lime whitewash) were expensive, only the front fence would be given that coating. If they were painted at all, a cheaper color, usually

Finishes for fences range from traditional paints and stains to newer, clear protective coatings.

red, was applied to the side and back fences. The same economizing was also applied to the house itself.

Painting, at least until the introduction of pressure-treated lumber, was the only practical method for protecting fences against the ravages of the weather. Wood that is weathering is wood that is disintegrating. Although most woods, cedar and redwood above all, age with grace and beauty, the poetic filigree of fissures and raised grain is also a landscape of harsh and relentless erosion. The question isn't whether to paint or not to paint, but what to apply to the wood that will preserve both its life and its beauty.

"We have tried many kinds of washes, but found nothing that is not mixed in oil, that will endure much better than common simple lime wash." *(The Country Gentleman, 1862)*

Until the advent of acrylic latex paints, lime whitewash was the only tried and true water-based paint. Any of several recipes from 1930s-era USDA publications would be useful to anyone whose do-it-yourself conviction extended to mixing up their own fence paint. Here's one for "a more permanent whitewash":

• Dissolve 12 lb. of salt, ½ lb. of powdered alum, and 1¼ lb. of sugar in 10 qt. of hot water.

• Mix thoroughly 50 lb. of hydrated lime in 15 gal. of hot water.

• Combine the two solutions. This mix makes about 20 gal., more than enough to whitewash the average fence.

The modern fence builder can choose between an increasingly varied selection of traditional paints and stains and new water-repellent clear protective finishes. There are benefits and drawbacks to each.

PAINTS

All modern paints intended for outdoor ("exterior") use contain preservatives, fungicides, mildewicides (which is why they shouldn't be used on "interior" surfaces), insect and water repellents, as well as a plethora of additives formulated to prolong the life of the coating and impart desirable physical properties, such as ultraviolet resistance. (The high energy of ultraviolet radiation breaks the molecular bonds in the paint film, making it lose its grip and turning clear coatings cloudy.)

Up until the 1980s, when advances in paint chemistry improved the properties of water-based paints, professional painters and experts alike agreed that oil-based paints were superior for outdoor use. However, the debate may well be settled by default. There are signs that the necessarily high "volatile organic compounds" (VOC) content of oil-based paints and solvents will lead to a general ban on their further use. (The hazardous fumes of VOCs are a major cause of smog.)

There's still a convincing argument (for as long as it lasts) that oil-based primers provide more "tooth" than water-based ones since they penetrate deeper into the wood fiber. Only oil-based primers are effective at preventing the natural aromatic extractives in redwood and cedar from bleeding through the finish coat. Oil-based top coats are arguably regarded as more durable than modern latex, but they are also considerably more expensive.

Although water-based top coats can be applied over an oil-based prime coat, the reverse is not possible. While this might seem to settle the question insofar as primers are concerned, premium acrylic latex primers provide a base that is actually more flexible and thus more resistant to thermal stress than the harder oil-based coat. Unlike oil paints, a latex paint film is permeable to water-vapor migration, which otherwise lifts the film, causing blistering. Latex paint is also easier to clean up and, with a much lower VOC, less hazardous both to living creatures and the environment. The recent introduction of even less toxic "green" paints that rely on citrus-based emulsifiers instead of petrochemical polymers means that environmentalists can have their fence and paint it, too.

With the availability of treated lumber and a variety of clear protective finishes, the choice whether to paint or stain is largely one of fashion, not function. A painted fence also ties in better with a painted house. Paints also last longer and tend to look better than stains on the smooth surfaces of dressed boards. Like the heavy makeup models wear for photo shoots, a couple of coats

In Colonial times, fences were typically painted to match the color of the house.

Painting the fence and house in contrasting colors can be very pleasing as long as the colors are selected with care.

of paint can mask the defects in low-grade wood so it looks good, at least from a distance.

STAINS AND CLEAR COATINGS

Because the ratio of vehicle (the solvent) to the pigment that floats in it is so much greater in stains than in paints, stains penetrate deeper into the wood and are more readily absorbed into the fibers. Albeit tough, a paint film adheres only to the surface, which is why it degrades with age, flaking, and peeling; stains weather, fading gracefully away. And, as a bonus that alumni of the T. Sawyer school of fence

painting would appreciate, since no primer or undercoat is necessary, stains are faster, easier, and less costly to apply than paints.

Stains can be clear, semi-transparent, or "solid color" (opaque), depending on the ratio of pigment to carrier. Clear (nonpigmented or only slightly tinted) stains enhance the natural color of the wood as well as the pattern of its grain and knots. Clear (or slightly tinted) coatings are also vehicles for the various silicone-based water-repellent and preservative sealers that retain the natural look of the wood while protecting it against ultraviolet radiation and moisture-related weathering. It's a good idea to use a sealer as an

undercoat for the primer on very porous woods.

Because pigments also endow the coating with its durability, clear stains and water-repellent finishes are short-lived. Depending on the product and the severity of its exposure, recoating at six-month (in regions of heavy rainfall) to two-year intervals is part of routine maintenance. Nevertheless, the benefits of the protection afforded are worth the cost and effort. One way to reduce the frequency of recoating is to use one of the recently introduced pressure-treated lumber products that have also been treated with a water repellent.

Don't confuse clear finishes with varnishes, urethanes, and other color-

Hardly traditional, this purple picket fence is eye-catchingly delightful.

less resin-based coatings that are completely unsuited for outdoor use. Urethane varnish has so little ultraviolet resistance that it will cloud, yellow, and flake away in less than a year. Even spar varnish, which is specially formulated for exterior use, offers poor long-term performance.

Semi-transparent stains contain enough pigment to color the wood, but not so much that the grain and other natural features are obscured. These stains are a good choice for "pre-weathering" new wood to match the old. Since any stain or clear finish will change the appearance of the wood and the same stain will look different on different kinds of wood or in different kinds of light, experiment on test pieces first before applying any of these products to the fence proper.

In addition to color and surface · sheen (i.e., flat, semi-gloss or satin, and gloss), check that the products are compatible not only with each other but also with the type of wood they'll be applied to. Cedar and redwood, for example, react unfavorably with many coatings; not all stains are suitable for use with pressure-treated wood, while certain water-repellent sealers are specifically formulated for immediate application to it. Cedar and redwood also stain in contact with metals, requiring isolation by caulk, paint, or rubber gaskets.

Solid-color stains are more like thin paints. They contain enough pigment to hide the grain, but not enough to mask defects, since they don't actually form a surface film. Although less durable and more easily marred than a paint film, opaque stains weather much more slowly than lighter stains. The paintlike color density of solid color stains makes them a good choice for dressed lumber, and an excellent, less labor intensive substitute for outdoor paint in general. Unfortunately, water-based stains do not perform as well as oil-based stains. However, the current generation of oil-based stains are now "VOC compliant" (meaning that they meet the latest state air quality standards).

Bleaching oils and bleaching stains are used to soften the rawness of new wood or to blend repairs more quickly into the background. By reacting chemically with the wood and the elements, bleaching oils quickly give wood the weathered gray tone that it would otherwise take two or three years to acquire. There are also bleaching formulations that will restore weathered gray wood to not quite its original but still very rich tone. Most of these compounds contain harsh chlorine bleaches or acids that are harmful to plants and animals and hazardous to use. Newer, environmentally benign products include granular powders that are added to water to make a thick foam that is brushed onto the fence to "brighten" it.

Paint coatings require tiresome upkeep and eventual recoating to maintain their protective benefits.

five

PLANNING & LAYOUT

The boundaries of the estate are made more secure

by the building of fences, which prevent the

servants from quarreling with the neighbors, and make it

unnecessary to fix the boundaries by lawsuits.

Marcus Terentius Varro (116–27 B.C.)

Planning involves more than drawing a quick sketch of your fence and ordering some materials. For most fences, you'll need to develop a site plan, establish your fence line, and make an accurate layout on the ground. But even before taking these steps, you'll first want to check that your proposed fence meets local codes and ordinances.

Good neighbors make good fences. Think about the social and legal aspects of your fence before you build it.

LEGAL MATTERS

As explained in Chapter 3, fence building in urban and suburban areas is a highly regulated activity, subject to a variety of local statutes, subdivision or planned-community rules, and deed restrictions. In rural areas, fence building is less strictly regulated, but it's still a good idea to discuss your proposed fence with your neighbors.

In our litigious society, building all or part of your fence on your neighbors' property can be grounds for a costly lawsuit. At the very least it makes for ill will. It's only common sense to ascertain the exact location of your property lines even before you look into any applicable fence laws. A copy of the most recent official "plat," or survey map, for your property will be on file in the registry of deeds at your town, city, or county clerk's or assessor's office.

Deeds and maps are often ambiguous, however, and it's not unusual for the exact location of a boundary line to be unclear. If the actual boundary line cannot be located, the law allows the adjoining property owners to decide where they want it to be. This requires each of the owners to sign a quit-claim deed that legally describes the new boundary. The deeds are then recorded at the town office to make them valid and transferable with the property. Be sure to check with your mortgage company before entering into any such agreement with your neighbor.

Since the local laws governing the building and maintenance of boundary-line fences can be quite complicated and are not at all uniform, it's always a good idea to check with town officials

or even a lawyer for any potential legal problems that building your proposed fence may present.

DEVELOPING A SITE PLAN

Early on in the planning stage, you'll need to get some idea of how the fence will fit into your property. This process usually begins with a site plan.

The site plan should be something more than the minimal schematic that accompanies the building permit. Start by making a rough sketch of your property (or the relevant portion of your site if the area to be fenced off comprises only a fraction of your property). The sketch should also show the house, patios, decks, sheds and outbuildings, swimming pools, driveways and walkways, large trees or other plantings, garden and service or play areas, and any other salient features that might figure in the fence layout (see the drawing on p. 122).

Next, using a 100-ft. tape and a helper to stretch it tight, take measurements of the perimeter and of the other prominent features and record them on the sketch. Transfer these measurements to a 17-in. by 22-in. sheet of ¼-in. graph paper. (The dimensions of even a small lot won't fit onto a standard 8½-in. by 11-in. sheet at ¼-in.-per-foot scale.) The predrawn squares on the graph paper make it easy both to scale off dimensions and to draw parallel and perpendicular lines without a drawing board and other drafting tools.

Locate all the features you've recorded. Use a #4 pencil and a light touch for your preliminary work, and then go over the pencil lines with a fine-point nylon-tip marker. The pencil lines and any smudges can be removed with a gum eraser after the ink dries.

Now comes the tricky part: figuring out how to reconcile the givens of the site with the requirements of what an architect would call the "program," that is, the purposes you want your fence to fulfill. Begin by listing the general priorities for the fence, or fences, such as boundary, privacy, security, ornament, screening wind or noise, that will correlate to specific areas on the plan.

When planning a garden fence, make sure the gate opening is wide enough to allow a wheelbarrow to pass through. (Photo by Bill Rooney.)

Sample Site Plan

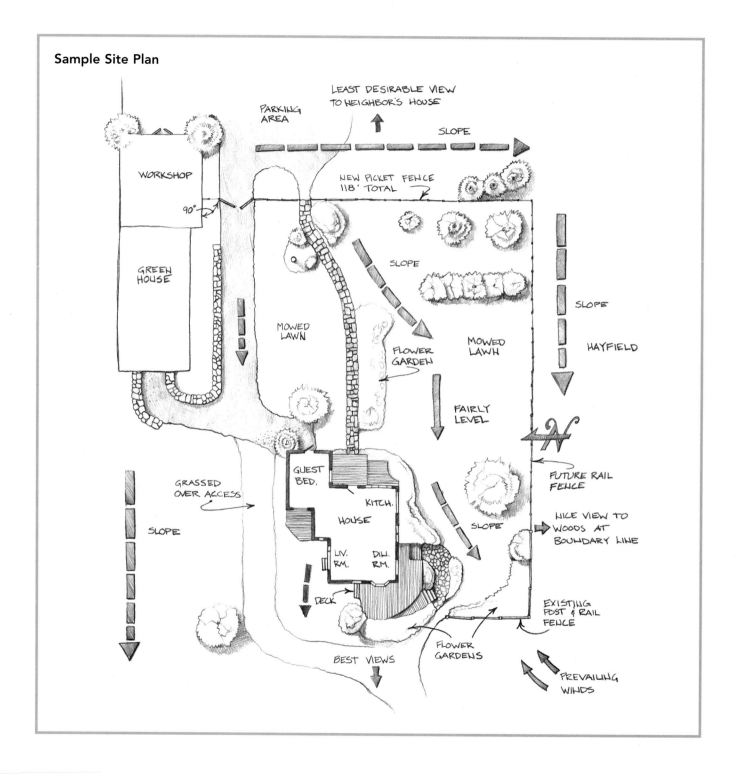

LEAST DESIRABLE VIEW
TO NEIGHBOR'S HOUSE

PARKING
AREA

SLOPE

WORKSHOP

90°

NEW PICKET FENCE
118' TOTAL

GREEN
HOUSE

SLOPE

SLOPE

MOWED
LAWN

HAYFIELD

MOWED
LAWN

FLOWER
GARDEN

FAIRLY
LEVEL

GRASSED
OVER ACCESS

GUEST
BED.

KITCH.

HOUSE

FUTURE RAIL
FENCE

NICE VIEW TO
WOODS AT
BOUNDARY LINE

SLOPE

LIV.
RM.

DIN.
RM.

SLOPE

DECK

EXISTING
POST & RAIL
FENCE

BEST VIEWS

FLOWER
GARDENS

PREVAILING
WINDS

Think about the immutable aspects of your site, things like where the sun rises and sets, the direction of the prevailing winds and the sources of noise that you might wish to mitigate, where the good and bad views are, and features on your site that you might wish to conceal, remove, improve, or retain. Also consider the various discrete activity areas such as the garden, children's play area, pet area, space for outdoor entertaining or lawn games, and the traffic patterns that run from the street to the house and that connect one area or feature to another.

The relationship between activity areas and traffic patterns determines where to locate your gates. The type of traffic determines their width. As a general guide, a 3-ft.-wide opening will allow one person and small equipment like lawn mowers and rototillers or wheelbarrows to pass through (see the photo on p. 121). While this might be fine for a side or service entry, a 4-ft. minimum is a better choice, particularly for a formal front gate. Obviously, you'll need a wider gate to admit a car or pickup truck comfortably. If you have any doubts about how wide the gate opening should be, make a full-size mockup by marking the proposed gateposts with a pair of trash cans, sawhorses, or traffic safety cones, and then try driving between them.

Don't forget to allow for any future remodeling or landscaping plans, such as a swimming pool, an addition, or any other changes that call for moving heavy equipment onto your site. It's easier to provide a gate now rather than tearing down sections of fence later. Another option is to build an easily removable section of infill that's wide enough for machinery to pass through.

The next step is to locate or otherwise indicate these program priorities on a site-plan overlay. This exercise is intended to help you see how these factors might affect the placement of your fence (refer to the drawing on the facing page).

Once you've collated all this information in graphic form, you can begin making a tentative fence layout. As you do, try to imagine how the fence might appear from inside the house, from your yard, and from beyond your property. (Showing the location of your home's windows and doors and the room layout on the initial

When developing the site plan, try to imagine how the fence will look—both from the house and from beyond your property.

site plan or the program overlay helps facilitate this process.) Use tracing paper overlays to try different layouts without regard to redrafting the original plans.

Once you've decided on a final version of the layout, draw it carefully on a fresh sheet of tracing paper. Since you've already worked out all the details on the preliminary overlays, there's no need to include them on this working drawing. Just indicate the outlines of the house and other significant features. But since this is the plan that you'll use to figure your materials and the spacing between the fence posts and the location of the gates, it should be carefully drawn and accurately dimensioned.

The foregoing discussion more or less assumes a fence that's part of a major landscaping project involving the entire property, as might be the case when building a new house or when making extensive site renovations. Frankly, most fence projects are seldom so complicated or extensive that they require the sort of multifaceted planning detailed here. The location of the average fence is more likely to be one of the givens, as, for example, when it runs along the perimeter of the property or marks a single boundary line. Even with a simple fence, however, you should still, at the very least, make an accurate working drawing. Not only will you need one

to estimate materials but also working out the actual details of construction on paper rather than in wood and nails can save wasted time and materials.

ESTABLISHING A FENCE LINE

The majority of all fences, certainly at least those built in the cities and suburbs, are boundary fences. The first step in laying out a boundary fence is to establish the fence line. Usually this is easy to do because the iron pins that mark the corners of your property are still visible or there is some otherwise clearly established legal point of reference from which you can take your bearings. Stretching a string between two pins in the ground is simple. So is setting up a string parallel to an existing known line such as a sidewalk or an alleyway. But if your initial search for the property markers is unsuccessful, you'll have no choice but to hire a surveyor to reestablish the property lines.

LAYING OUT A STRAIGHT FENCE

Once you've located the property markers, set a pair of stakes about 1 ft. back from the pins (as a precaution against any inadvertent incursion on your neighbors' land) and stretch a string between them to mark your actual fence line. If it so happens that your fence line is unrelated to your property line, its layout will usually correspond to some other given of your

property. Most fences typically run parallel with or at right angles to references such as a driveway or walkway, the line of a building, landscape features like ditches, or the margin of a lawn or woods. Here, you can simply set your stakes and string wherever you decide you want the fence to be.

Well, actually, it's usually a little more complicated than that. Assume that your lot (or at least the portion of it along the fence line) is basically flat and level and that your fence is a simple straight line, with no corners or curves. Even though it's then possible to lay out your posthole centers from this initial string (see pp. 138-139), it would be a lot more helpful if the string also established a level reference line for setting stringers or rails and post heights, and if you could remove it while digging the postholes and put it back up to set the posts in a straight line.

This is what batter boards are for. A few feet beyond each terminus of your fence line, drive a pair of 2x4 stakes firmly into the ground so that they straddle the line by 1½ ft. to 2 ft. on either side (see the photo at left on p. 126). Since you don't want the stakes to shift, brace them diagonally laterally and from front to back. Next, level and fasten a straight-edged 1x6 board across the first pair of stakes at the height you wish to set your string. The trick is transferring this reference level point to the other batter board.

A straight fence on a level surface is the easiest type of fence to lay out.

The starting point for the layout of the author's fence was a post set against the wall of an existing workshop.

The first step in laying out the fence line is to stretch a string between batter boards at either end of the fence.

There are several ways to transfer the reference points. A suggestion that appears in many do-it-yourself books is to use a line level, which is essentially a leveling bubble encapsulated in a short tube. In theory, you hook this device onto the middle of a tightly stretched string fixed at one end to your reference point and held at the other end by an assistant, who moves it up and down against the other batter-board stake until you see the bubble read level. At this point, your assistant marks the intersection of the string with the stake, and you level across to the other stake and attach the second batter board and tie the string to it. In real life, it's almost impossible to stretch a string of any significant length tight enough so that it won't sag and throw off the measurement. Furthermore, because the line level is only 3 in. or 4 in. long, misreading the position of the bubble by the merest hair translates to several inches of error at the far end of the string.

A better alternative is to use a water level, which is one tool that can't be anything but absolutely accurate. Although you can buy water levels, it's pretty easy to make your own, as explained in the sidebar on the facing page.

MAKING AND USING A WATER LEVEL

You can make a water level from a garden hose fitted with a 1-ft.-long piece of rigid clear plastic tubing at each end, or you can simply buy 50 ft. of ¼-in. flexible clear plastic tubing.

Fill the hose with water, either with a funnel or by siphoning (which is a lot easier). To siphon the water, put one end of the tube into a pail of water (add food coloring to make it visible in the level) and suck on the other end. All the air bubbles must be removed or the reading won't be correct. To test the level, bring both ends next to each other. If the water levels are different, you've still got air in the line. Refill and try again.

To use the level, hold one end so that the surface of the water in the tube (the meniscus) lines up with your reference mark. Since water always seeks its level, as your helper moves the tube at the other end of the hose up and down against the stake, the meniscus in your end will rise and fall. When it lines up with your original reference mark, signal your helper to mark the level of the meniscus on the new post. This mark represents the height at which you should fasten the second reference board.

Using a Water Level to Establish a Level Line

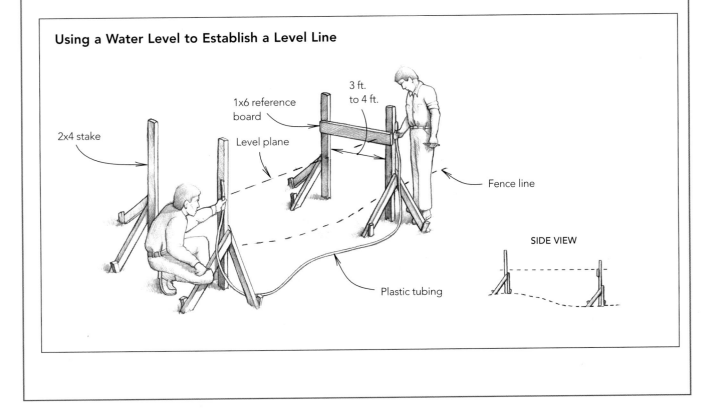

2x4 stake

1x6 reference board

Level plane

3 ft. to 4 ft.

Fence line

Plastic tubing

SIDE VIEW

A sight level is a simple tool for establishing a level reference line.

for establishing the reference line. Unless you're a professional builder with enough reason to own one, this instrument is one tool you'll want to rent rather than own. If you've never used one before, a transit level can appear pretty intimidating. Fortunately, using one is a lot easier than it looks. Essentially, the only difference between the transit level and the sight level is that the scope is held level by a tripod instead of your hand. Just the same, it's a delicate piece of machinery that must be handled carefully in order to maintain its accuracy. Be certain the level you rent has been calibrated, and get expert advice on how to use it.

Another way to establish a level reference line is to use a sight level, which is a simple tool that will give accurate readings ($+/-\frac{1}{4}$ in.) over 100 ft. or so. It's basically a small (about 6-in.-long) handheld telescope fitted with a leveling bubble. To use the level you'll need a helper at the far end of the line. Hold the level against your benchmark and sight into it until you can see the tip of the pencil your helper is moving up and down the other stake line up with the internal crosshairs of the scope. At this point, signal to your helper to mark the stake. Sight levels are relatively inexpensive (about $25) and simple to use.

When the fence line involves right angles or other angles, long lines of sight, or obstructions in the line of sight, a transit level is the tool of choice

LAYING OUT A CORNER

Laying out a corner is not that much more difficult than laying out a straight line. You can use a transit level to set the right angle, but if you don't feel comfortable using this tool, the traditional 3-4-5 triangle method is just as effective.

To run a true 90° corner, set a stake under your baseline string at the intended corner of the angle. Stretch a second string from the center of the corner stake to a batter board (or stake) at the far end of the new fence line (see the drawing on p. 130). Then hook your tape measure over a nail in the corner stake and mark off a distance on the baseline string corresponding to a multiple of 4 (20 ft., for example). Repeat the process on the second string, this time marking off a multiple of 3 (15 ft., for example).

A right-angled corner is easy to lay out using the 3-4-5 triangle method.

It follows from the rules of geometry that any triangle whose sides are a multiple of 3, 4, and 5 must be a right triangle. Since the chances for error increase with the length of the line relative to the triangle it's projected from, if the site allows, I'd use a 15-20-25 or even a 30-40-50 multiple.

Although you could manage it working alone, the next step is much easier with a couple of helpers (especially with a big triangle). Have

the first assistant hold the "dumb end" of a long tape directly over the mark on the baseline. The second assistant moves the end of the second string along the batter board until its mark coincides with the length of the third multiple on the smart end of the tape measure that you are holding, at which point he or she fixes the string on the batter board.

You can extrapolate from the diagonal to lay out other angles as well. Set batter boards and, with your helpers, stretch a string diagonally over

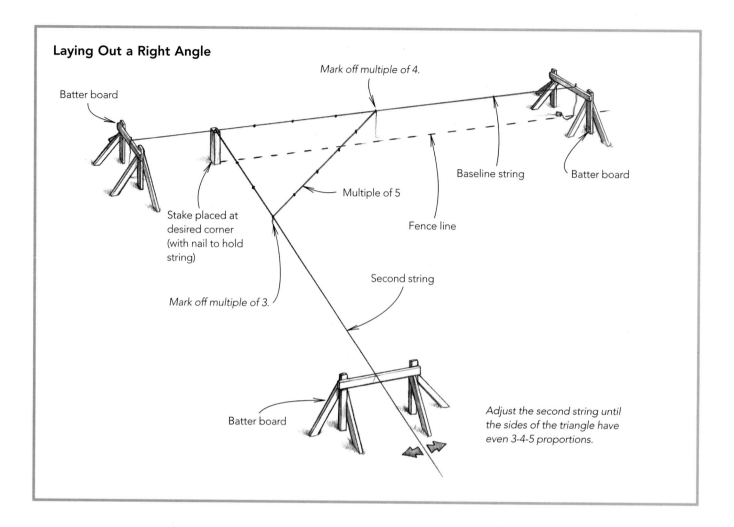

Laying Out a Right Angle

Mark off multiple of 4.

Batter board

Stake placed at
desired corner
(with nail to hold
string)

Mark off multiple of 3.

Multiple of 5

Baseline string

Batter board

Fence line

Second string

Batter board

*Adjust the second string until
the sides of the triangle have
even 3-4-5 proportions.*

the measuring marks. Then mark off the diagonal at the distance corresponding to the desired angle. Bisecting the diagonal yields a 45° angle, trisecting it describes 30° or 60° angles. After you've run and set a fourth string from the corner stake through the section point, the right angle and diagonal strings can be removed.

LAYING OUT A CURVED FENCE
Not all fences run in straight lines or turn sharp corners; some prefer to curve. There are two approaches to laying out a curved fence. You can make a true arc or you can approximate an arc with a series of short chords. Which technique you choose depends largely on whether the infill will consist

of individual boards or pickets or sectional panels.

Laying out a true curve is a prerequisite for establishing chords in any case. All you need is a very large compass. First, set stakes to mark the ends of the proposed curve on your baseline (see the drawing on p. 132). Then locate and mark the midpoint

STRINGERS FOR CURVED FENCES

Laying out a curved fence is a lot easier than building one. The problem lies in making the curved stringers that support the infill.

You have two choices when it comes to making curved stringers: laminate them "on edge" or laminate them "on flat." The on-edge treatment is the easiest.

On a gentle curve, screw the end of a strip of ½-in.-thick bender-board (or a 1x4) long enough to span two or more fence bays to the starting post. With a helper holding the far end, bend and screw the board against the next posts. Trim the end to break on the last post center before screwing it home.

After the first layer is installed on all the posts, install a second layer, offsetting the joints by one post bay. Glue the laminations together with outdoor construction adhesive, clamping as necessary until the glue sets. If desired, strengthen the joint with screws.

To bend laminated stringers in a severe arc, decrease the thickness and increase the number of plies. Steaming the boards to soften them for tighter bending is also an option.

The process of making on-flat laminations consists of inscribing segmented arcs into chords of short lengths of stringer stock and cutting them on the bandsaw. As with on-edge lamination, stagger the joints between segments.

A curved corner is an interesting alternative to a sharp, right-angled corner.

between them. Next, run a string at a right angle to this midpoint. It should be at least as long as the distance between the end stakes. Now drive a 3-ft. or 4-ft. length of iron rod (pipe or rebar) into the ground along the new line to serve as the pivot point of your compass. The distance between the pivot and the center stake determines the curvature of the arc: the closer the pivot, the steeper the curve, the farther out, the shallower.

Tie a piece of string to the pivot point and secure the other end to a sharp stake or a chalk bottle at the point where it intersects the end stake of the curve. Keep the string taut as you scribe or chalk the arc on the ground. You may have to experiment with the pivot-point location to get the desired curve. Once this is done, lay a flexible tape along the arc to subdivide it into even increments for the posthole centers. These same points also establish the chords for a sectional curve.

To lay out an undulating fence, repeat the layout on alternating sides of the baseline.

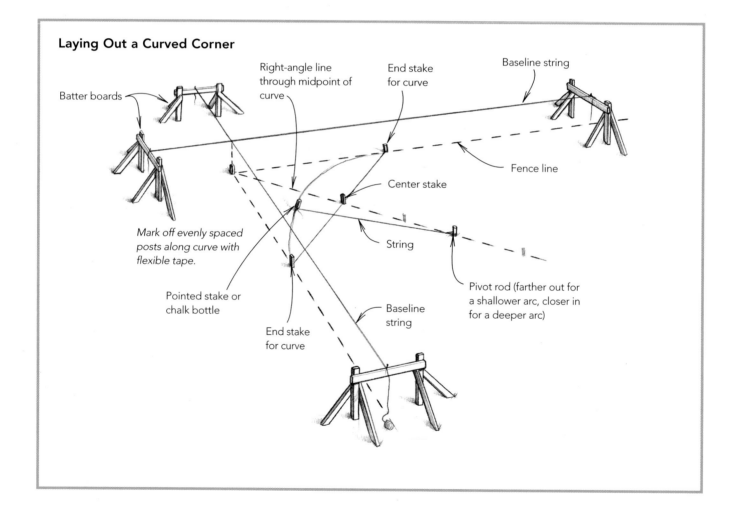

Laying Out a Curved Corner

Batter boards

Right-angle line through midpoint of curve

End stake for curve

Baseline string

Fence line

Center stake

String

Mark off evenly spaced posts along curve with flexible tape.

Pivot rod (farther out for a shallower arc, closer in for a deeper arc)

Pointed stake or chalk bottle

Baseline string

End stake for curve

LAYING OUT A FENCE ON UNEVEN TERRAIN

For the sake of simplicity, thus far we've been pretending to lay out a fence line on more-or-less level ground. Now it's time to deal with the techniques for sighting around the obstructions and negotiating the slopes of real estate that has some terrain to it.

Sometimes you may have to establish a fence line to a point that isn't visible from the starting point. You'll need a few helpers to do this. Start by driving stakes at both end points. Then stand at the first stake, have a helper stand at the hidden stake, and position two more helpers holding tall poles in between. The tops of both poles must be visible to each observer.

Sight from your position and motion to your pole people, shifting their positions until their poles are in your line of sight. Then the hidden observer directs the pole holders until they are in line with his or her point of view. Continue alternating sightlines and repositioning the pole people until they appear to be in both observers' lines of sight. Set stakes at each pole position and run a string between all four points.

The technique just described works only when there aren't any obstructions between the end points of the fence line. When something large and immovable like a building, boulders, a body of water, or unfenceable thicket prevents you from running a line from A to B, you can establish a parallel line where your line of sight is clear and measure back from it to locate the end points of the line segments on each side of the obstruction.

LAYING OUT A FENCE ON A SLOPE

As explained in Chapter 3, if your fence is to be located on a slope, you can either let the fence follow the contour of the terrain or step it down the slope in level panels, with stringers set level between the posts. If you know how much your land slopes from one end of the fence to the other, you can

One way for a fence to negotiate a slope is to have it step down the slope in level panels.

decide which is best suited to your terrain. Knowing the slope also lets you calculate the number of steps (when the bays are narrower than your standard bay) and how high each one should be to achieve the most balanced appearance.

Calculating slope To calculate the amount of slope (or, as carpenters call it, "pitch"), you need to find the overall height differential from one end of the fence line to the other. If you're using a transit level, the difference in height is simply the difference between the benchmark reading and the end-point reading on the target rod. Divide this figure by the number of fence-panel

sections to get the height of each step, assuming you want all steps to be equal.

On a relatively constant slope, simply brace a plumb stake on the downslope of the fence line. Rest the end of a straight 2x4x12 on the upslope, and have a helper raise and lower the other end against the stake until the bubble centers in the level you're holding on top of the 2x. The ratio of the difference between the level line and the ground (in feet) and the length of the straightedge is the slope (for example, 2 in 12).

You can also use levels and strings to measure a slope. On a relatively gentle slope, set a short stake in the ground at

the top of the slope. Set a stake whose top is at least slightly higher than the short stake at the low end of the slope. Tie a string to the upper stake at ground level and stretch it across the end of the longest hand level you can manage to hold against the face of the lower stake. Move the line up or down until the string reads level. Mark the stake and measure to the ground to find the rise of your slope. Measure from the short stake to the edge of the long stake to find its run. Use these numbers to plot your slope on graph paper when you design the fence.

On a steep slope, a level line starting at grade level might be 20 ft. or 30 ft.

LEVEL STRINGS AND LINE STRINGS

It's important, at least for a conceptual understanding, to distinguish between fence-*level* strings and fence-*line* strings, even though the same string generally performs both layout functions. Besides measuring slope and marking panel steps, level strings are used to mark a level cut-off line at the top of the fence posts and/or the level point where the horizontal rails or stringers are attached to the posts. Although it

can be, a fence-line string doesn't have to be level. Its function is to represent the line ("run") of the fence and to keep the posts and rails in line as they are installed.

A fence line also serves as a kind of level line when it's used to indicate the top of a contour fence that follows uneven terrain. To maintain the requisite, more-or-less constant difference between the string and the grade, set plumbed and braced grade stakes at inter-

vals along the line (to keep the line straight, set them a hair away from the string). After all the stakes are set, drive nails into them at the constant height and mark and set the string on top of them. You could also dispense with the stakes entirely and simply set your posts untrimmed and overlong down the line. The string then marks the cut-off and top rail lines.

When a fence starts at the wall of a building, it's most pleasing if it runs out at a right angle.

above grade at the end of the fence. It would take a stake as high as a tree and a very tall stepladder to level the string. Here, you'll have to use a transit level, or extrapolate the slope on paper.

Stepping a fence down a slope To illustrate how to lay out a fence that steps down a slope, I'll use the example of the picket fence featured in this chapter and the next. The particular problems I had to solve in the layout of the fence exemplify some pertinent general design considerations for fences on irregular slopes.

There were two critical site givens that dictated the layout of the fence. First, one terminus of the fence was against the wall of an existing workshop (see the photo above). Running the fence line into the strong line of the building at anything but a precise right angle would appear awkward and disconcerting. Second, although there was a height difference of 8 ft. over the fence's 118-ft. run, the slope was quite

The author checks to make sure the fence-line string lines up with the end of the walkway.

irregular, with one short, steeply dropping section, several longer gentle slopes, and a stretch of near level ground, all of which precluded even steps and initially suggested a design that mirrored the contour of the ground. Another complication was the existing walkway of flat stones set in gravel with 4x4 pressure-treated margins (see the bottom photo on p. 135). The end of the walkway was intended to coincide with the opening for the fence gate.

I began by stretching a string from the far end of the walkway border to the shop wall, shifting it from side to side until it lined up with the edge of the 4x4. I marked that point on the wall and extended it about 3 ft. up onto the wall with a level, which I had already decided would be the approximate

height of my fence. I then tied a string to a nail at this point (see the bottom photo on p. 125) and ran it several feet past the far end of the fence line, tied the end off to a stake, and drove it into the ground. Next, I straddled the string with a batter board set at the same height off the ground as the starting point. I transferred the string to the top of the board, adjusted its position until the line ran even with the end of the walkway, and checked to see that it was perpendicular to the shop wall.

Looking at the line approximating the top of my fence, I realized that following the contour would be a big mistake. The problem was that an angled fence line doesn't read very well against the strong level line of the

house. So I decided to make the fence step gradually down the slope in level panels (see the photo on p. 123).

With the fence line established, the next step was to figure out the number, place, and height of the panel steps. Since the gateposts were fixed points, they were a good place to begin. Having already measured and plotted the slope and potential post layout on paper, the posts would fall on roughly 5-ft. centers (see the drawing below). Although I had originally proposed a 12-ft. driveway gate, shrinking it down to 11 ft. allowed the three panels between the entry gate and the driveway gatepost to observe the 5-ft. centering, at the same time leaving more than enough clearance for the fuel-oil truck without deviating

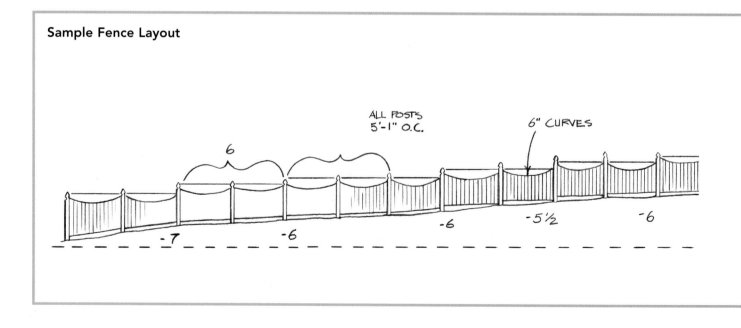

Sample Fence Layout

ALL POSTS 5'-1" O.C.

6" CURVES

6

-7   -6   -6   -5½   -6

noticeably from the "5-to-8 rule" panel proportion (see the sidebar on p. 67). The interval between the fixed walkway gatepost and the driveway gatepost was conveniently divisible into approximate 5-ft. intervals. And because the far terminus of the fence line was somewhat arbitrary, the same 5-ft. spacing module worked out evenly on the other side of the gate as well.

I set story-pole stakes at each gatepost and at various multiples of 5-ft. intervals along the fence line wherever the contour changed distinctly (see the photo at right). I set my sight level against the benchmark on the first story pole and read to the next, where my daughter marked the measurement. Dropping down the story pole an amount equal to the difference between

With the fence line established, the next step is to figure out the number, place, and height of the panel steps.

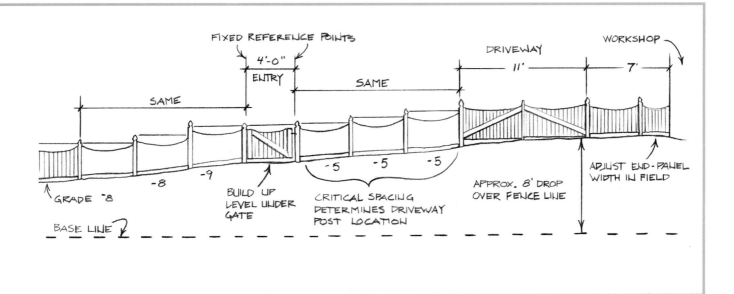

To lay out the posts on level terrain, transfer the post-center location from the site plan to the string.

Use a level to plumb down from the post-center marks on the string to the ground.

the first and second readings, I sighted over to the next pole.

I repeated the operation along the entire fence line, writing the height differentials on their respective story poles so that I could transfer them to my scale drawing. Then, by drawing in the fence-post locations, I could divide each major differential into even steps according to the number of panels it encompassed. Fiddling around with the differentials and the location of the steps smoothed them out so that no one step was noticeably bigger than any other. To get an idea of how the fence might look in real life, I repositioned the string on the story poles to show the proposed steps. This engendered another round of fine-tuning adjustments until my wife and I were both satisfied with the profile of the fence.

LAYING OUT THE POSTS

Whether you're laying out a fence on a slope or on level ground, once the baseline string is established, it's time to measure and mark the location of the fence posts. On level ground this is simple. Referring to your site plan, mark the location of the post centers on the fence-level string using a tape measure hooked to the beginning post and an indelible marker (see the top photo at left). Plumb down from the spacing marks to the ground with a 4-ft. (or even 6-ft.) carpenter's level, as shown in the bottom photo at left. Then drive a 20d nail or larger spike through a 3-in. by 3-in. square of cloth, plastic, or heavy paper—any bright and water-resistant material will do—to mark the centers.

A time-saving alternative to the mark-and-plumb method is to mark off the post centers with a gauge pole laid alongside a fence-line string stretched just above the ground. Select a straight length of 1x2 to use as a gauge pole and cut it to the exact distance between the post centers. Drive marker pins into the ground where the corner of the gauge pole contacts the string. Pins are more precise than stakes and easier to set and remove. If you aren't quite sure how well your post spacings will transfer from your drawing to the field, do a test layout first. Just push the pins into the ground by hand instead of hammering them home. Then you can check to see how your fence sections will end against fixed points like gate and terminal posts and adjust the spacings for the most balanced look.

Because the hypotenuse of a right triangle (the ground) is always longer than its side (the level fence line), the gauge-pole-and-pin method may not work as well on steeply sloping ground. This is particularly the case with stepped infill. Here, to maintain the correct spacing, the post centers must be dropped down from the level line as explained previously.

In the case of my fence, a test measurement with a straightedge and level resting on the ground indicated that the difference in length between

To calculate post spacing on a steep slope, level a gauge pole from one marker pin (left) and then hold a level plumb against the string to locate the coordinates for the next pin (below).

level and slope was negligible, except at three rather steep dropoffs in the grade. (In any case, since the postholes are two or three times wider than the post, there's room to correct any small errors in centering.) So, after removing the story poles and running the string in a straight line from one end to the other, I hooked the end of my tape on a nail driven into the center point of the first post and stretched it along the length of the fence line. I marked the center

Precise post spacing is critical for picket fences to ensure that the picket spacing module remains constant.

points of the posts on the string with an indelible marker at the calculated 5-ft. spacing. Then, holding my level plumb and just barely off the string, I set the center pins in the ground.

On a steep slope, a better approach is to calculate the post spacing from a level line drawn on paper and then set your fence line at a convenient height above grade. Tape a 4-ft. level to the top edge of your gauge pole and lay it on the ground with one end on the first posthole-center pin (see the photo at left on p. 139). Raise the opposite end

of the gauge pole until it reads level (it helps to have an assistant for this part), and then use another 4-ft. level held plumb against the string and the end of the gauge pole to locate both coordinates for the next center pin. Repeat on down the line.

POST SIZE AND SPACING
For most fences, the spacing interval of the fence posts is determined by the width of the infill panels, which in turn is decided according to aesthetic and structural criteria. I've already explored

RECOMMENDED POST SIZE AND SPACING

To ensure the structural viability of your fence, use the following guidelines for post size and spacing:

- 4x4 posts on 6-ft. centers work well with light or heavy infills and low fences, or with light infills and tall fences.
- 4x4 posts on 8-ft. centers work for light infill and slightly heavier infills with rails on edge. They are also the minimum required for board fencing (two to five boards).
- 4x4 posts on 10-ft. centers mark the upper limit of low full-inch-thick horizontal board fencing (two to three boards). They are acceptable for wire-mesh garden fences or rail fences with toe-nailed or notched-in peeled saplings (not to exceed 3½ in. diameter).
- 6x6 posts on 8-ft. centers are structurally and visually strong.

They are good for heavy board or plank infills, both vertically and horizontally, and for through-mortised peeled-pole rails. Use 2x4, 2x6, or 4x4 rails.
- 6x6 posts on 10-ft. centers will easily carry heavy infills and rails for both low and tall fences. Use 2x6 or 4x6 rails. The criteria for choosing 8-ft. or 10-ft. spacing are visual rather than structural.

Six-by-six posts on 8-ft. centers are suitable for heavy board or plank infills.

Gateposts can
be sized larger
than the line posts
(above), or all
the posts can be
oversized to
maintain visual
continuity (right).

the aesthetic considerations in Chapter 3. At this point, it's enough to reiterate that together with the frame, fence posts are a major visual component of the fence design. Even when hidden from view by an uninterrupted infill on the front side of the fence, the visual rhythm and balance or the lack thereof in the post spacing will be obvious on the back side of the fence that you'll be looking at every day. The only exception is a fence whose posts and structural frame components are completely concealed by the infill on both sides.

That said, there are certain structural considerations that also govern the sizing and spacing of fence posts (see the sidebar on p. 141). The basic principle is that the posts must be thick enough and close enough together to bear the weight of the infill they carry. Thus, the taller the fence, the heavier (and the deeper) the post. Gateposts, which must bear a greater load than line posts, and terminal posts, which aren't braced by the rest of the fence, also follow this rule. (For more on anchoring gateposts, see pp. 189-191.) The normal practice is to dimension these posts one size larger than the line posts, which perforce requires a deeper footing. You could also oversize all the posts to maintain visual continuity (see the photos on the facing page).

Remember, also, that as you increase the post spacings, the stringers must be wider to prevent sagging. Eight-foot centers are generally considered the upper limit for 2x4 rails. In my opinion, using 2x4s for spacings much over 6 ft. is pushing it, especially with a heavy infill. I'd consider using 2x6s instead as low-cost insurance against sagging. They're definitely a requirement for any fence rail in excess of 8 ft.

six

BUILDING A FENCE

Those who build most expensively do not

necessarily secure the most tasteful

places, and in fencing there is much opportunity

to let thought balance money.

Frank J. Scott, *The Art of Beautifying Suburban Home Grounds* (1886)

All fences, no matter how simple or elaborate, no matter the style or materials, are ultimately three-part structural systems. While the actual forms or construction details of each of the three basic components—the footings, the frame, and the filler—are subject to myriad variations, all fulfill the same functions and contain the same subsystems.

The footings do for the fence what a foundation does for a house: They anchor the frame securely in the earth, keep the fence standing plumb and upright, and impart the stability it needs to withstand the pressures of wind, animals, and climbing children. A proper footing consists of the post-hole itself, the drain bed that supports the post and allows water to percolate away from its base, and the concrete, gravel, or native soil backfill that packs the hole and actually anchors the frame.

The framework is a two-part component consisting of vertical posts and horizontal stringers, which together comprise the structure that carries the infill. The posts join the fence to the ground. The stringers join the posts to each other. Sometimes, as with a post-and-rail fence or post-and-board fence, the stringers are also the filler (in which case, they are known simply as "rails"). The filler (or infill, to be more precise but less alliterative) is the visually prominent surface that we usually perceive as the fence proper.

In this chapter, I'll show you exactly how this structural system goes together, from digging the postholes to adding the final decorative flourishes. The fence featured in this chapter is a traditional picket fence, but the techniques illustrated apply to most other fence types as well. At the end of the chapter, I'll focus on some details that can make a stronger fence and on some ways to dress up a plain fence.

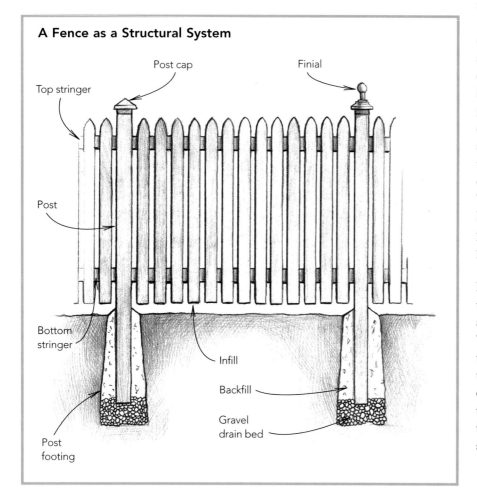

A Fence as a Structural System

Top stringer

Post cap

Finial

Post

Bottom stringer

Infill

Backfill

Gravel drain bed

Post footing

DIGGING POSTHOLES

Most books about fence building are either strangely silent or breezily glib about the subject of digging postholes. For example, even as the authors of the mid-19th-century *Village and Farm Cottages* advise the reader to "let the hole be of the smallest possible diameter, and twice as deep as frost ever reaches," they admit, perhaps somewhat disingenuously, that "there is some trouble in this, but it pays in the end."

The nature of your soil determines how hard you'll have to work to dig your postholes. The fact that, except for a brief experience with the concrete-hard caliche of Arizona, all the post-holes I've ever dug have been in the frost-prone hardpan and bony clay of Vermont and Maine probably explains my jaundiced view of the enterprise.

HOW DEEP SHOULD THE HOLE BE?

The traditional rule of thumb has always been to set fence posts at least one-third of their length: hence, 2 ft. for a 6-ft. post, leaving 4 ft. above grade; 32 in. for an 8-footer, with just over 5 ft. above grade; and at least 3 ft. for a 10-ft. post and a 6-ft. exposure. By this standard, the post for a 3-ft. fence would be buried only 18 in. deep. Since I don't recommend setting any post less than 2 ft. deep, I'd dig the extra 6 in. in order to anchor the post as securely as practical.

Because they must provide extra support, gate, corner, and terminal posts (the posts at the beginning and

Sometimes the digging is easy, sometimes it isn't. All these stones were pried and pinched out of a single posthole.

end of a fence line) should be set at least 1 ft. deeper than the line posts. And whenever a fence or post must withstand greater than normal stress, such as high winds, very loose, soft, or wet soils, or a heavy infill, the standard practice is to set the posts as deep as possible. In any case, since regional practices generally reflect regional realities, it's always a good idea to ask a local fence contractor about standard practices for anchoring fence posts in your area.

If and when soil conditions and the capabilities of your digging equipment permit, it certainly won't hurt and definitely will increase your peace of mind to sink your postholes beneath the frost line. The depth of the average frost line shown on maps or specified by your local building code is calculated for bare ground. In actuality, the probability and severity of frost heave doesn't depend on temperature as much as it does on soil conditions, which are very much site-specific. Frost heave can occur only when water is held in the soil. So sandy, gravelly, and other well-drained soils aren't as likely to be plagued by frost heave.

On the other hand, heavy clay, silt, and muck soils that hold water very well or even more porous soils saturated by a high water table will heave as much as an inch for every foot of penetration. But since a blanket of uncompacted snow is a surprisingly effective insulator, your fence posts won't necessarily heave out of the ground if they aren't set below the frost line (as long as the snow cover lasts all winter).

Postholes filled with soil should be at least twice as large in cross section as the post. Also, postholes should widen out at their bottoms: Ideally, the bottom third of the hole should be at least a third wider than its top. If the hole is wide enough, chip away at the sides with a spade shovel held backwards to enlarge the bottom. Otherwise, work at it with a clamshell digger. When backfilled with concrete or other material, the anchoring effect of the resulting bell-like footing will be similar to one of those punching dummies that you can't knock down.

DIGGING BY HAND

The traditional way to dig postholes is by hand, using shovels and a clamshell digger. The main problem with digging postholes by hand is that the deeper you dig, the wider the top of the hole has to become. Otherwise, you can't tilt the shovel handle far enough back to scoop the dirt out of the hole. Also, as the hole gets deeper, the shovel handle effectively shortens, until you lose your leverage.

Hand-dug postholes basically bottom out at about 3 ft. Fortunately, in most parts of the country, and for most fence posts, this is more than deep enough to get below the frost line and to anchor gate and line posts up to 8 ft. tall above grade. Where it isn't, you'll either have to live with the possibility of heaving posts or find a way to dig a deeper hole.

Where the soil is soft enough, you can gain an extra 6 in. to at most 1 ft.

Hand tools for digging postholes include shovels and a clamshell digger. The author uses the levels and story pole when setting the posts.

by switching over to a clamshell digger. You can also borrow a trick from Mother Nature and fill the holes with water, letting them soak until the bottom softens. Then bail or pump out the mud until you hit hard ground. The only other way to go deeper by hand is to dig a hole big enough to stand in while you dig. I'd recommend hiring a backhoe before resorting to this option.

If you are lucky enough to live on land that lets you dig holes with a hand shovel, you can probably skip the next section. The rest of you should read on.

DIGGING WITH AN AUGER

The next step up from digging post-holes by hand is to use a power-driven posthole auger. There are three kinds: one- or two-person handheld augers, tractor-mounted augers, and commercially operated truck-mounted augers. Which type you resort to depends on how many holes you need to drill and on the obstinacy of the soil you need to drill them in.

Handheld augers Using a handheld gas-powered auger, you can drill a post-hole that would take you half an hour or longer to dig with a shovel in less than a

minute. Two-person machines typically rent for about $45 a day, which is plenty of time to dig at least 150 holes. But speed of digging isn't the same as ease of digging. Drilling even 20 or 30 holes is exhausting work, even in good soil, since all that keeps the auger's power head from spinning itself and anyone attached to it around is the weight and muscle of the operators. When the auger lodges against a rock, the sudden shock to your already stretched tendons can be brutal. Don't let this possibility tempt you to rent the cheaper one-person machine—it simply doesn't have enough power for the job.

As with any construction project that involves excavation, be sure to check for the location of any underground utility lines before digging your postholes. If you can find where an underground line penetrates the house foundation or floor slab, you can generally assume that it runs in a more-or-less straight line directly to the utility pole, curbside transformer, well shaft, or water, gas, or sewer line.

If you must dig without knowing the exact depth and location of an underground line, proceed carefully with a hand shovel. Be on the lookout for a change in soil texture as you dig: Underground lines are typically embedded in a protective buffer of stone-free soil or sand before burial. Never use a power-driven posthole auger in a suspect zone. And remember, too, if you're digging where there's only a little room for error, there's a lot bigger chance of making one.

Fortunately, although the line of a fence is continuous, the postholes themselves are relatively small and few and far between. The likelihood of digging into a buried line is actually quite remote. The chances of seriously damaging one are even smaller, at least with hand tools. (This doesn't apply to the soft copper tubing used for buried LP gas lines.) Polyethylene water pipe and direct-burial underground service cable are both tough enough to deflect a hand shovel. So is the PVC plastic pipe now used instead of galvanized steel and cast iron for municipal water and sewer lines. And in many locales, underground service cable is run through protective PVC conduit. Telephone cable is pretty tough, too, but you can drive a shovel blade through it.

Hold that digger!—at least until you've checked for any buried utility lines.

The key to safe and trouble-free operation of a handheld auger is not to run the auger so fast that you lose control of the machine nor so slowly that it binds. Never attempt to drill straight to the bottom of the hole in one shot. The auger can bury itself before you realize it and stall out. Instead, clear the hole by backing the auger out every so often as you drill.

Encountering even a fairly small rock can make the auger veer off plumb and skew the bottom of the hole, necessitating a lot of remedial spade work. When you dig out the rock and try to rebore the hole, the auger will keep jumping back into the loose dirt in the old one. Another shortcoming with rental machines is that they are usually equipped with only 3-ft.-long, 6-in.- or 8-in.-diameter augers. While 3 ft. are certainly better than none, you've got a problem when that's not deep or wide enough for your posts.

Tractor-mounted augers The next level of escalation is to use an auger mounted on the three-point hitch of a farm tractor and powered by its "power take off" (PTO). Depending on the capacity of the auger and the power of the tractor, the limitations of this machine are basically similar to those of its gas-powered little brother. The main advantage of this type of auger is that it substitutes the tractor's horsepower for your muscle power, which is no small thing when you have a lot of postholes to dig. (However, a tractor-mounted auger won't digest stones any better than a handheld auger.) Some of the more rugged units will drive a 12-in. or even a 16-in. auger with a 4-ft. to 6-ft. cut, but most of the machines you can rent are set up for 8-in. augers with a 3-ft. maximum depth.

A tractor-mounted auger takes two to operate: one person sitting on the tractor to manipulate the hydraulic controls that raise and lower the machine and to engage and disengage the PTO drive; and the other to set the auger on the mark and steady it while it's drilling (see the photo at right). Tractor-powered augers cost about $25 an hour to rent.

Truck-mounted augers Seriously rocky and difficult soils call for heavy artillery. You can hire a truck-mounted hydraulic auger that's used to drill the holes for setting utility poles, complete with the operator, for about $400 per day. Some operators will charge a minimum rate for small jobs, which may be

A tractor-mounted auger speeds up the process of digging holes considerably.

less than the daily rate, or else quote a fixed price. Since a truck-mounted auger can drill upward of 200 holes in a day, the cost per hole is reasonable when you've got a lot to dig. It's also reasonable when you've got only a few dozen holes to dig in ground where nothing short of a rock drill will work.

REMOVING ROCKS

Even the power of a truck-mounted auger is no match for solid rock. What are your options when one or more posts are located over a really big boulder or even solid rock that goes all the way down to the continental plate? The first step is to determine the extent of the problem.

If your posthole auger snags momentarily and then just spins in place, use a clamshell digger to clean out the bottom of the hole and feel around with a digging bar (a heavy 6-ft.-long crowbar with a flattened end rather

than a point) or, space permitting, a spade shovel to find the edges of the stone. Then see if you can catch an edge and wiggle the rock. As a rule of thumb, if you can move it, you can remove it. The sound your bar makes when you drive it hard against the stone is also a clue to its movability: A dull thunk indicates a fairly small rock; anything too big to move by hand will emit a sharper, more ringing tone when struck.

Stones up to the size of a grapefruit can be removed with a clamshell digger. For larger rocks, you'll have to enlarge the hole to dig around them.

Usually you can extricate a rock from the side of a hole by digging down along its edge until you can undermine it and lever it out. When the rock covers part of the bottom of the hole, use your clamshell digger to dig out the unobstructed area. If you're lucky, you'll be able to dig past the bottom of the stone, undermine it, and pry it free.

The clamshell digger is also handy for enlarging the bottom of the hole to facilitate the removal operation. Hold the jaws open but parallel to each other and jam one blade sideways into the hole. Once the hole is sufficiently enlarged, you may able to speed things

Some rocks are better left alone— unless you have a backhoe, you could spend the better part of a day digging to remove a bother- some boulder.

Pinning the posts to an existing brick wall is one way to avoid the onerous task of digging postholes.

up with your spade. Don't expect rapid progress—it can take an hour or more to dig out a volleyball-sized boulder.

If your investigation reveals a rock too big to move by hand (for me that's generally a rock that doesn't even budge when I can get a solid purchase under an edge with my crowbar), you have two realistic choices: leave it there or bring in a backhoe to remove it. Unless the hole is for a gatepost, if it's at least 2 ft. deep, I'd leave it alone and plan on pouring concrete around the post when I set it. If you can't dig down 2 ft. and the rock is immovable by any means, you'll have to relocate the fence post. One final option for the fanatical is to drill a hole in the rock and join the post to it with a length of rebar (as explained in the sidebar at right).

ANCHORING THE POSTS

All the trials and tribulations of digging postholes will come to naught if the posts aren't solidly anchored. People sit on, lean against, and climb fences. Horses push against them. Snow piles up and wind blows against them. In wet ground, fence posts lean drunkenly. Frozen ground wants to spit them out. Insufficient resistance to lateral loads is the death of fences.

If you think of a fence post as a lever, then it's obvious why the first line of defense is to set the posts at the "correct" depth. Despite all the rules of thumb and depth-to-length formulas, the correct depth all too often turns out

---

PINNING A POST TO A ROCK

If your fence post can't be relocated to avoid an immovable rock or a solid ledge, the next best option is to drill a hole in the rock and pin the post to it. For a 4x4 post, use ½-in. rebar; for a 6x6, use either ⅝-in. or ¾-in. bar. Standard carbide-tipped masonry bits will work if the rock is no denser than brick, which, unfortunately, is not likely to be the case with the boulder at the bottom of your posthole. It takes a heavy-duty electric hammer drill and an impact bit to drill even a small hole in dense rock like granite.

You can also drill the hole by hand, albeit slowly, with a star drill (a foot-long hardened steel bar with a fluted point that looks like the tip of a Phillips-head screwdriver). Give the drill a quarter-turn between each blow from a 2-lb. sledgehammer. Always wear safety glasses to protect your eyes from flying stone chips and metal shrapnel.

Once you've drilled the hole in the rock, scribe and trim the bottom of the post to fit the rock and hammer the pin into the hole. Set the post on the pin, check for correct position, and mark and drill the hole in the post. Coat the bottom of the post with asphalt cement and drive it onto the pin. Plumb and brace the post and fill the hole with enough concrete to cover the rock and at least 6 in. of the post.

---

to be only as deep as you can dig. If you can dig deeper than the 2-ft. minimum, by all means do. The deeper you can set your post relative to its length, the shorter the handle of the lever that's the above-ground portion becomes. When soil conditions make digging even to minimum depth impractical, you must devise a way to compensate for the longer lever.

Under ideal conditions, the bottom of the posthole would be at least 1 ft. below the frost line. In practice, however, it's usually more practical to set your posts in shallower holes and backfill them in a way that minimizes the opportunities for frost penetration. Depending on regional custom, professional fence builders will advocate for the superiority of concrete backfill or tamped earth and gravel with equal passion. Barring any local code restrictions, the choice is up to you.

CONCRETE FOOTINGS

It's a common notion that filling the hole with concrete protects a post from frost heave. It can't, and it doesn't. Frost will heave buildings and tear down mountains. If it can get under the bottom of a post, it won't have any trouble popping a tiny cork of concrete out of its hole. What concrete does do, however, is anchor a post when the soil alone doesn't provide enough resistance to lateral or uplifting forces. Until it settles, the backfill around a newly set post is fairly plastic, which means that it's quite easy to push the post from side to side. But since the undisturbed soil surrounding the hole is much denser, filling the hole with concrete increases the stability of the post.

The brunt of frost action on a fence post tends to be within the ring of frozen ground in the first few inches immediately below the surface. If the post is anchored in a solid plug of concrete, the frozen earth above it counteracts the uplifting grip of the

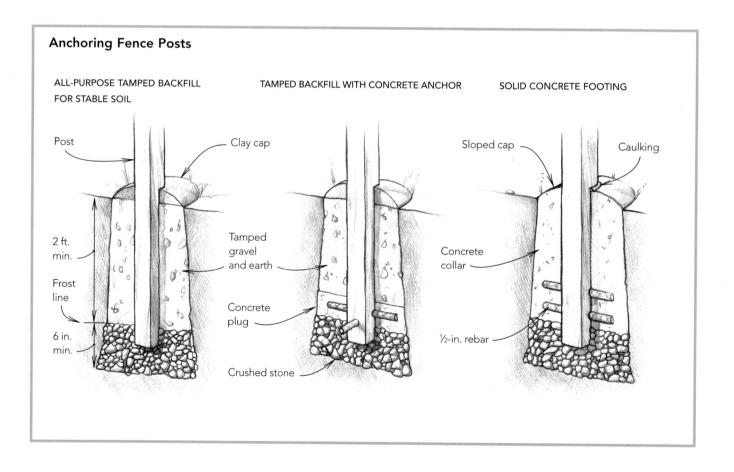

Anchoring Fence Posts

ALL-PURPOSE TAMPED BACKFILL FOR STABLE SOIL

Post

Clay cap

2 ft. min.

Frost line

6 in. min.

TAMPED BACKFILL WITH CONCRETE ANCHOR

Tamped gravel and earth

Concrete plug

Crushed stone

SOLID CONCRETE FOOTING

Sloped cap

Caulking

Concrete collar

½-in. rebar

frost on the post. On the other hand, unless it goes deeper than the frost does, a concrete collar filling a posthole from top to bottom can be even more vulnerable to frost action than a post set in soil alone.

There are other problems with continuous concrete collar footings besides ineffectiveness against frost heave. Some fence builders mistakenly believe that setting untreated wood in concrete protects it from decay. In fact, the opposite is true. A wood post, especially if it's green, will shrink away from the concrete. The resulting gap is an ideal environment for decay and insect attack. Also, water trapped in the crack will freeze and expand, causing the concrete to break. To forestall this disaster, use seasoned lumber for the posts (even if using pressure-treated wood) and seal any cracks with a high-grade flexible and durable caulk or hot tar as soon as they appear. Also, slope the top of a freshly poured footing away from the post to help direct water away from the joint (see the drawing on the facing page).

It's important that the bottom of the post not be embedded in the concrete. Instead, it should extend at least 2 in. into the crushed-stone base. Any water that gets between the post and the concrete can then drain harmlessly away.

Another potentially troublesome aspect of encasing posts in concrete is that you'll need to dig wider postholes. A hole filled with concrete must be at least three times larger than its post (as opposed to two times larger for a hole filled with earth). The reason for this becomes clear when you consider that a 4x4 post in the middle of an 8-in. column is surrounded by only 2¼ in. of concrete. Such a thin cylinder has little of the resistance to lateral forces that is the rationale for using concrete in the first place. The practical consequence is that you'll need at least a 12-in. auger to bore the holes for 4x4 posts when they're set in concrete.

Since most picket fences, or fences with thin horizontal boards, peeled poles, or split rails, offer minimal wind

Half-inch steel rebar will help anchor this post to its concrete footing.

ONE BAG OF CONCRETE PER HOLE

If you have a lot of concrete to pour into a lot of holes, it makes sense to rent an electric mixer and to buy sand, crushed stone, and 90-lb. sacks of portland cement to batch-mix concrete.

If you have only a few holes to fill, purchasing the concrete premixed is more convenient. An 80-lb. bag of premixed concrete (such as Sakrete, a trademark that's generic in lumberyard-speak) will yield approximately 1 cu. ft., or enough to surround a 4x4 post set in an 8-in. to 10-in. hole with about 16 in. to 18 in. of concrete.

A wheelbarrow easily accommodates a single bag of premixed concrete.

resistance, setting their posts in concrete is structural overkill (which doesn't mean it's a bad idea.) However, the posts for a typical tall residential privacy fence, whose large and solid surface area is the structural equivalent of a sail, always require extra strong anchorage.

TAMPED-EARTH FOOTINGS

The question of whether or not to use concrete depends largely on the stability of the native soil. If you're setting posts for a low fence in dry soil that compacts well, you could probably fill the holes with the same dirt that came out of them. Note that tamping the soil firmly in 2-in. or 3-in. layers as you fill up the hole will anchor the post much more solidly than filling to the top and tamping once.

Embedding a post in alternating layers of tamped earth and gravel is a centuries-old method for firmly anchoring fence posts that works best in stable soils, somewhat less satisfactorily in wet clay soils prone to severe frost heave, and not very well at all in sand. The gravel provides drainage, while the packed earth provides lateral stiffness. A variation that amounts to the same thing is to mix the gravel and soil into a fairly porous and firm amalgam.

In areas where digging becomes impossible well above the frost line, you can backfill the hole with pure gravel to create an isolation zone between the post and frost-prone

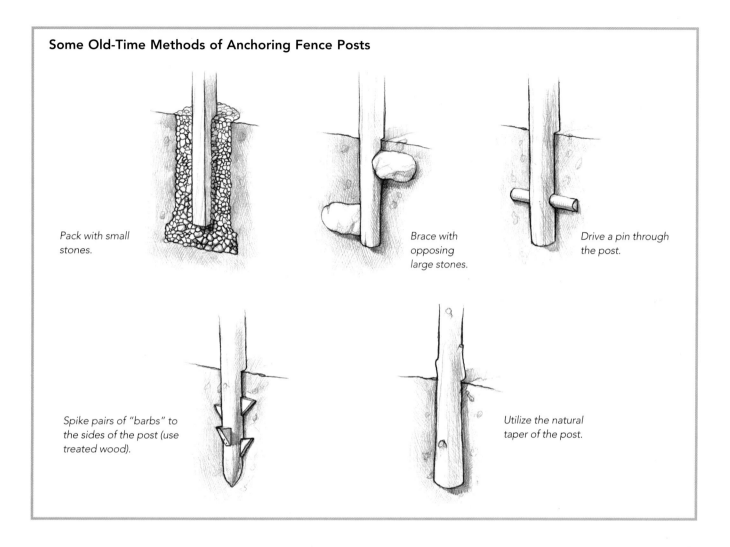

Some Old-Time Methods of Anchoring Fence Posts

Pack with small stones.

Brace with opposing large stones.

Drive a pin through the post.

Spike pairs of "barbs" to the sides of the post (use treated wood).

Utilize the natural taper of the post.

surrounding soil. Any ground water will move rapidly down through the gravel and seep into the subsoil at the interface with the hardpan layer. Capping the gravel with a protective layer of tamped earth, shaped to drain water away from the post, will help prevent surface water from waterlogging the hole.

The heavier the soil, the more clay it contains and the poorer it drains. When clay becomes wet, it basically liquefies and looses its stability. Under these conditions, even a deeply planted fence post won't have much lateral strength. There's no reason why some of the methods old-time farmers used to keep their fence posts firmly planted in soft ground won't work just as well in a modern suburban backyard (see the drawing on p. 157).

Tamped gravel or crushed stone belongs at the bottom of every posthole, regardless of soil type. Ground water can move through this porous base and percolate into the surrounding soil instead of saturating the backfill in the hole. This reduces the potential for frost heave and retards the progress of decay. After you've widened and cleaned out the bottom of the hole, fill it with 4 in. to 6 in. of crushed stone or washed gravel and compact it firmly with a tamping bar. If you have a digging bar,

The standard approach is to set overlong posts and then trim them to height after they are backfilled.

you've already got a tamping bar since it's the belled head on the other end. A sledgehammer, a capped length of pipe, or even a 2x4 also makes a satisfactory posthole tamper.

SETTING THE POSTS

There are two basic ways to set the fence posts. The standard approach is to set overlong posts, plumb and center them on the fence line, add diagonal braces in two planes to maintain alignment, and then backfill. With the posts set, you can use a level line to mark the tops for cutting, and then take measurements off that line to find the location of the stringer notches. Next you cut the notches in the field, that is, on the already installed posts. With stepped fences, you simply transfer the level from section to section, either with a straightedge or a water level.

The second approach, which is a little more complicated, is to set posts that have already been cut to length and notched for the stringers. This is usually how the traditional post-and-rail fencing that graces so many suburban landscapes is installed. There are two ways to deal with precut posts. If the fence follows the contour of the grade instead of true level, all the postholes must be the same depth. Dig the posts about 6 in. deeper than the full length of the below-ground portion of the post. Make a combination depth-gauge and tamping tool from a 2x4 with a pair of braced crosspieces bolted to it at the post-depth mark. Fill and tamp the bottom of the hole with gravel until it reaches the correct level.

When the rails are to be set level, the bottoms of all the postholes must likewise be level. Reset the level string (which is why you should leave batter boards up until the fence is finished) to establish a constant distance with regard to the level of the gravel at the bottom of the postholes. Draw a corresponding mark across the top end of your tamping rod so that it becomes a story pole to indicate when you've reached the desired level.

SETTING THE POSTS FOR A PICKET FENCE

The fence I designed and built for this book called for a pair of level stringers to be set in notches (or "dadoes") cut in the posts. In addition to the notches, the posts terminated in an ornamental flourish, or finial, which began at a precise distance above the upper notch. I decided that it would be easier to precut this post cap before setting the posts (see p. 175).

In itself, the design of the fence posed no more installation problems than a standard through-mortised post-and-rail fence. The complication was due to the difficulty of digging postholes to uniform, let alone adequate, depth in my stony soil. Where the soil was cooperative, I went down about 3½ ft.; where it wasn't, I stopped wherever the rocks told me to. The standard method of filling the postholes

to a constant depth with tamped gravel was out of the question.

Since the decorative bottom veining of the finial was to be 9 in. above the upper stringer notch, the requisite adjustment in post length to ensure level infill panels would have to be trimmed off the bottoms of the posts *before* the notches were marked and cut.

To build this kind of fence, the first step is to check that the starting post-hole is where it's supposed to be. Drop the post in the hole, hold its face against the fence-line string, and quickly check for plumb. (Since this string would have originally been used to mark off the posthole centers, it should be shifted to mark the post faces. If you prefer, offset the string to allow for a gauge block as well.) Cut back the sides of the hole as needed so that there's

---

WORKING WITH GAUGE BLOCKS

One problem with using a fence-line string to guide your post and stringer installation is that a slight bow or twist in the lumber or a post that is somehow knocked off plumb after installation (which happens a lot more than you'd like) will throw off the line. The easiest way to avoid this situation is to borrow the technique carpenters use to "line" their walls. Nail 1x3 standoff blocks to the terminal posts and attach the line string to them. Use another piece of 1x3 for a gauge block to check the alignment between the string and the post or frame member.

Position an over-long post in the hole, roughly plumb it to the fence-line string, and then calculate how much to cut from the bottom of the post.

After cutting the post to length, plumb and brace it.

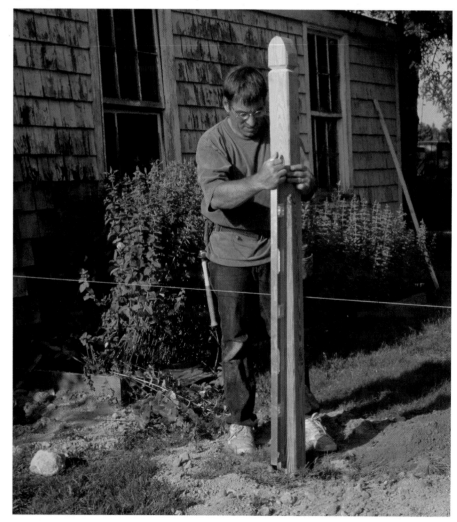

enough room for backfill on all sides of any off-centered posts.

Prepare the base of the hole by adding gravel and/or crushed stone and tamp it until it's smooth, hard, and as close to level as you can get it by eye. A level base minimizes the likelihood that any level line you mark on the post will shift relative to the reference string when the post is repositioned.

Next position an overlong post in the hole and plumb it to the fence-line string (see the photo at left). Now calculate how much needs to be cut from the bottom of the post. In the case of my fence, I measured the distance

1. Set a circular saw to the correct depth and angle and make the outline cuts of the notch. The square serves as a cutting guide.

2. Make repeated cuts with the saw to rough out the notch. This step requires care and an experienced hand.

3. Remove the waste and smooth the faces of the notch with a broad chisel.

4. Beveling the top of the notch and the stringer helps prevent decay by diverting water away from the backs of the pickets.

between the grade and the finial vein. Subtracting the desired height from this number gave me the length to cut from the post. Remove the post from the hole and trim the bottom to the calculated measurement.

Lay out the cut marks for the stringer notches. You can cut these in the shop with a radial-arm saw and dado blade or in the field with a circular saw, after the posts are set and backfilled (see the photos on p. 161).

Reset the trimmed post for the final time. Carefully pry the bottom of the post into exact position with the tip of your crowbar, plumbing and bracing it firmly in both directions (see the photo at right on p. 160). Backfill the hole in 3-in. layers, tamping each layer firm before shoveling in more fill (see the photo at left). End the tamped gravel about 2 in. below finish grade and cap the hole with heavy clay, sloped away from the post for good drainage.

Check the post once more for plumb. Don't despair if it somehow shifted during the backfilling. You can adjust a fresh-set post as much as ½ in. in any direction with a combination of a good sharp blow from a sledgehammer at ground level and some judicious tugging and pushing on its top.

To set subsequent posts, use a 1x3 spacing gauge with cleats attached to both ends to center each post off of its predecessor (see the photos on the facing page). For ease of attachment and removal, use screws rather than nails to fasten the spacing gauge to the posts. The gauge aligns the top of the new post so that it can be plumbed along the plane of the fence string. Level across from the stringer notches on the previously installed post with a straightedge or level to mark the notches on the subsequent post.

SETTING THE POSTS FOR A SECTIONAL FENCE

On a fence with notched posts like the one just described you can set all the posts first and then come back and

Backfill the hole with 3-in. layers of earth and gravel, tamping each layer as you go.

Working down the line, use a spacing gauge to center each post off of its predecessor.

install the infill from the back side. Mortised post-and-rail fences, on the other hand, must be assembled one section at a time. You can't set the posts in advance because the rails are longer than the fence bays and their tapered ends can't slide far enough by each other to let them slip between a pair of posts. Furthermore, if you're using peeled-pole rails with doweled ends, the mortises aren't cut all the way through the post. On these fences, once the first post is set, you'll need a helper to hold the free ends of the rails so you can fit the next post onto them.

If the convenience of setting mortised posts ahead of the rails is important, consider using a post-and-

Fences that use rails with doweled ends must be put together one section at a time.

On post-and-board fences, all the posts can be installed first and then the flat boards can be slid into the mortises.

board or ranch-style fence instead. The flat 1x boards (or 2x planks for sturdier fences) will slide past each other without difficulty. Farmers often took advantage of this detail to control access to little-used pastures in lieu of a hinged gate.

Certain styles of panel fences are also more easily assembled one section at a time. The frame of a store-bought or shop-built prefabricated infill panel can be used to space and align the posts. Once the first post has been plumbed and firmly anchored, run a fence-line string to a temporary terminal post. Trim the post tops to height before installing the panel (there may not be enough room to maneuver the saw afterwards). Mark the panel height on the post, allowing for sufficient clearance at ground level.

Attach a temporary diagonal brace to the post and lift the first panel into place, resting its bottom on blocking. If you're working without a helper, tack a stop (one or two short lengths of board will do) to the back of the post to keep the panel from toppling. Shim between the panel and blocking as needed to level the panel, and then screw the side frame of the panel to the post. If, as is typical with manufactured stockade-style panels, there is no side frame, consider adding one. Once the first panel is attached, position the second post against it, lined to the string and shimmed to level. Leave the diagonal braces in place

In some panel-fence systems, panels of horizontal infill are inserted into slotted posts. This gives the fence a clean uncluttered appearance and a very strong connection.

The best way to cut the slots in the posts is to use a table saw fitted with a dado blade. Since the slots should end at the bottom of the infill panel, the fence is laid out from the top down to leave extra length at the bottom ends of the posts for adjustment. The posts can be set in constant-depth post-holes or else be trimmed to length for each hole.

One advantage of this system is that all the posts are set first and the panels are slipped into the slots from above. This allows ample time for any concrete footings to cure and helps ensure a straighter fence run. On the other hand, since the slots allow only fractional end-to-end adjustments, the posts must be centered precisely.

until the fence is finished and all the postholes are backfilled.

Sometimes the absence of a side frame is actually an advantage. On an incline, for example, it's possible to rack a vertical infill panel to follow the slope. A typical 8-ft.-wide section of stockade-type privacy panel or 1x4 closed or spaced board or picket panels will rack up to 12 in. Up to 6 in. of adjustment is possible with 1x6 boarding.

INSTALLING THE INFILL

There are two ways to install the infill for a picket or board-type fence. The most common way is to attach the top and bottom stringers to the posts, and then nail on the pickets or boarding. The alternative is to prefabricate the fence panels and install them as a single unit.

BUILDING THE PANELS IN PLACE

Installing the stringers first and then the pickets (or boards) is the easiest and quickest method where the infill is continuous—that is, where it runs right over the face of the posts and stringers—but it can also be used where the infill is inset between the posts. All an inset infill requires is careful layout of the picket spacings and precise centering of the posts. Once the first picket or board is plumbed and attached, subsequent pickets can be quickly aligned with a spacing gauge made from a width of picket stock with cleats attached to hang it from the top edge of the stringer.

Choosing a picket pattern is part of the fun of designing your own fence. When my wife and I were planning our fence we knew that we wanted a fence that would be architecturally appropriate with our mid-19th-century Cape and one that would evoke a feeling of friendly reserve. We wanted something more than the plainsong of the square-top picket, something not quite as self-consciously cute as a popsicle-stick scalloped top, as pedestrian as the all-American diamond point, as effete as a filigree, and something less off-putting than the sharp arch of the true Gothic style.

I designed the picket tops by drawing a full-size pattern of the 1x3 picket stock, inscribing a semi-circular arc across its full width. I experimented with various versions of this basic arc, stretching it ever closer toward the sharp parabola of the Gothic arch. Once I had found a profile that we both liked, I copied the pattern onto ¼-in. hardboard for use as a cutting template. I cut the picket tops on the bandsaw, and then gang-sanded the tops to ensure a uniform profile.

1. Use a template to lay out the picket profile on the stock.

2. Cut the picket tops on a bandsaw.

3. Gang-sand the pickets to save time and to ensure a uniform profile.

PREFABRICATING THE PANELS

Although I could have built the infill panels for my fence in place, I decided that it would be easier to assemble the panels in my shop, using a jig to maintain a standard of sizing and spacing (see the photos on p. 168).

To build the panel jig, I stood a 4x8 sheet of ¾-in. plywood on its long edge and braced it to the shop floor. I drew lines to represent a full-sized pair of posts and the horizontal stringers. Because each panel of the fence was to have a downward arc rather than a level top, I draped a length of twine between two nails located at points representing the tops and centers of both posts to trace the curve of the picket tops.

Flipping the sheet horizontally on a pair of sawhorses, I cut the line with a jigsaw and then screwed a strip of ¼-in. plywood along the curve. This strip would serve as a stop for the picket tops. I nailed guide blocks to the plywood to space the pickets and stringers evenly and to hold them in

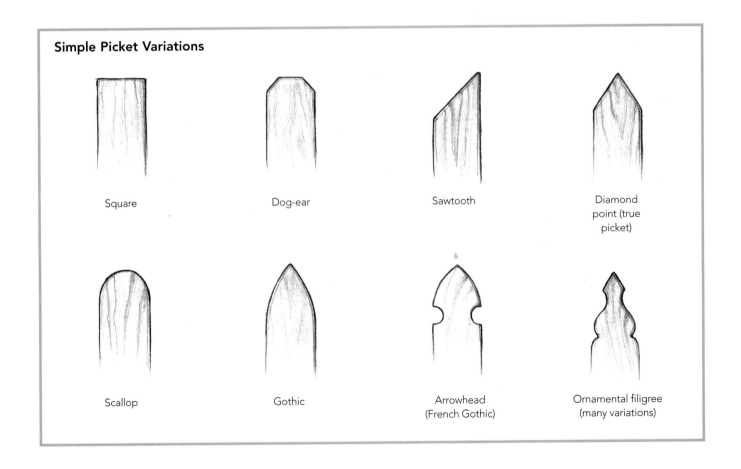

Simple Picket Variations

Square

Dog-ear

Sawtooth

Diamond point (true picket)

Scallop

Gothic

Arrowhead (French Gothic)

Ornamental filigree (many variations)

Using a jig for panel assembly ensures equal spacing of the pickets.

Once the pickets are nailed to the stringers, the panel is popped out of the jig and is ready to install.

position for assembly. Then I set the pickets in the jig and nailed them to the stringers, using stainless-steel 6d ring-shank nails. Allowing 2 in. for ground clearance, I snapped a chalkline across the bottom of the pickets and then trimmed the pickets to length. (On those panels that were to step down the grade, I measured between the underside of the lower stringer and the grade at both posts to find the slope of the ground and snapped a chalkline on the pickets to match.)

INSTALLING THE PANELS

Panels that fit into notched posts are easy to install. Simply slide the stringers into the notches from the back. Installing the panels for my fence was a little more complicated, however, because the tops of the notches were beveled to accommodate the beveled stringer (see the bottom right photo on p. 161). This meant that I had to slide them in from the side.

Installing the panels wasn't a problem on the flat stretches of the fence, where the stringers were only as long as the distance between post centers. But at each step where the stringers extended the full width of the post notches at both ends I discovered that I couldn't slide them into both notches unless I left off the last picket until the panel was in place.

After positioning the panel, check the posts for plumb again. Some of the stringers may have to be persuaded into place with a few taps of a sledgehammer. A tight fit is certainly aesthetically and structurally desirable, but the vertical alignment of the post can be a casualty of its attainment. As the number of installed panels accumulates, they tend to brace one another so that only the last post in the line will be out of plumb. Screwing the stringers to the abutting post first allows you to pull or push the offending post into proper alignment without dragging the stringer along with it.

After the panels are assembled, the stringers are slid into the mating notches in the posts.

The author's fence completed.

FENCE-BUILDING DETAILS

Clearly, the possible variations on fence styles are virtually unlimited. Although of a much lesser order of magnitude, even the specific practical details of fence construction present an overwhelming array of options. Design options have been presented in earlier chapters. Here, I offer a few more suggestions for making a fence structurally sound and durable and some simple techniques for adding

the decorative flourishes no properly turned out Victorian fence should be without.

DETAILS FOR A STRONGER FENCE

Stringers or fence rails add significantly to a fence's strength. They can be installed either flatwise or on edge. While there are many good reasons to install a rail flatwise, most of them have to do with the appearance of the fence, not its strength. Flatwise stringers are needed to carry double-sided or louvered infills or to keep the back side

of the frame flush with the face of the post. An on-flat stringer will strengthen the fence frame if it's laid over the tops of the posts. But even here, the relative vulnerability of wood laid flatwise to bending stresses means the stringer will tend to sag over long spans and heavy loads more readily than if it were laid on edge.

The type of stringer-to-post connection can also make a big difference in the stiffness and durability of the fence structure. Basically, the less the integrity of the joint depends on the

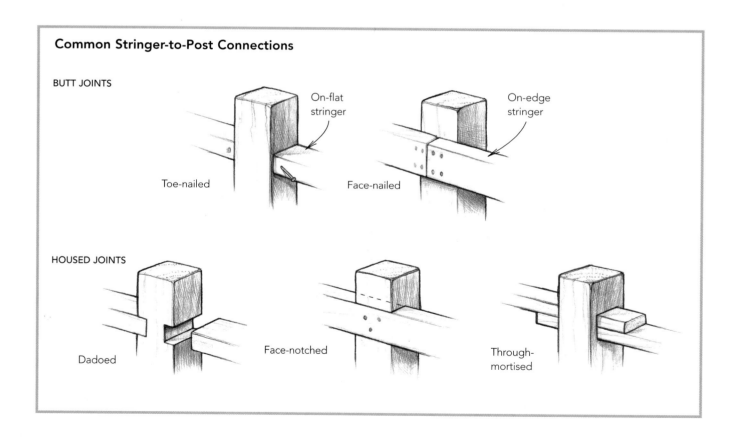

Common Stringer-to-Post Connections

BUTT JOINTS

On-flat stringer

Toe-nailed

On-edge stringer

Face-nailed

HOUSED JOINTS

Dadoed

Face-notched

Through-mortised

holding power of mechanical fasteners, the stronger that joint will be. The fact that centuries-old timber-framed houses are still standing today is proof of that principle. Although housed joints such as through-mortised, housed mortise-and-tenon, channel (rabbeted), and dado (notched or shouldered) joints are harder to make than butt or flush joints, the extra strength that they endow is almost always worth the trouble.

To cut a through mortise in a post, drill closely spaced holes along the perimeter of the mortise or bore full-width holes with a Forstner bit. Remove the waste wood with a chisel and then smooth the sides fair and true. To cut a slotted mortise instead of a square mortise, don't remove the material at the corners of the joint. Use a rasp file as necessary to smooth the sides or increase the radius of the slot.

PROTECTIVE DETAILS

Post caps and cap rails are more than ornaments. Any decorative detail that also diverts water away from the end grain of the posts and fence boards or that increases the structural strength of the framework can add years of life as well as beauty to the fence.

A column of wood is like a bundle of drinking straws. Any post top that will not be otherwise protected by an actual cap should be cut in some fashion to shed water rapidly so that it won't have a chance to be absorbed into the open pores of the end grain. Beveled,

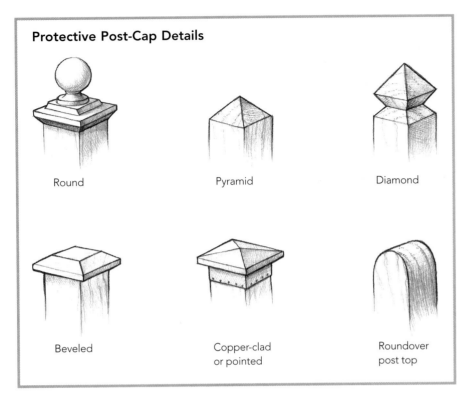

Protective Post-Cap Details

Round

Pyramid

Diamond

Beveled

Copper-clad or pointed

Roundover post top

Beveled cap rails and post caps shed water to protect the infill and the posts. (Photo by Bernard Levine.)

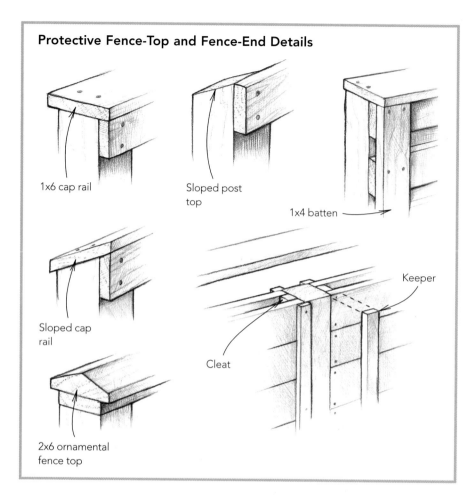

Protective Fence-Top and Fence-End Details

1x6 cap rail

Sloped post top

1x4 batten

Sloped cap rail

2x6 ornamental fence top

Keeper

Cleat

fences are likely to be climbed on or leaned against.

Battens and "keepers" both shield the ends of fence boards from the weather and help prevent them from pulling off the posts. They are especially useful for horizontal basket-weave fences and, when used in conjunction with a cap rail, will add a note of elegance and polish to any fence. Rails can also be detailed to help shed water by giving their top edges a 10° or 15° bevel.

Gravel boards, sometimes referred to as kickboards, protect vertical board infills by increasing the clearance between the ground and the bottom end grain of the boarding (see the drawing on the facing page). Once pressure-treated wood became available, routine replacement of rotted gravel boards was no longer necessary. However, where frost heave is normal even treated gravel boards should not be laid directly on the ground. If you fail to maintain the minimum 2-in. clearance, when the gravel board heaves it will take the entire fence, posts and all, along for the ride. One way to prevent this while closing off the gap under the fence is to leave the gravel board physically unattached to the bottom stringers and posts. Install keepers and arrange the infill so that the gravel board is free to slide past the face of the fence boards or into a hidden pocket beneath the bottom stringer.

Even without gravel boards, the bottoms of fence boards and pickets and stringers should be held at least

pyramidal, rounded over, and pointed arch cuts are all effective treatments (see the drawing on p. 171).

The umbrella of a cap rail will protect the infill (especially vertical boarding) as well as the posts against water. Cap rails also significantly increase the stiffness of the frame, which is particularly helpful when

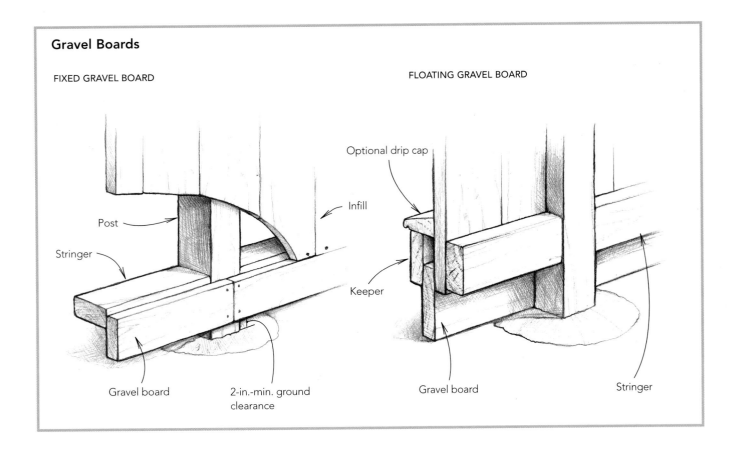

Gravel Boards

FIXED GRAVEL BOARD

FLOATING GRAVEL BOARD

Optional drip cap

Infill

Post

Keeper

Stringer

Gravel board

Gravel board

2-in.-min. ground clearance

Stringer

3 in. off the ground. Maintaining this clearance will protect the fence from ground water and, in cold climates, allow the frozen ground under the boards to heave without lifting the fence, too.

DECORATIVE DETAILS

The basic woodworking techniques described in this section can be used wedding-cake fashion to attain whatever level of ornamentation you desire. Although greater efficiency and precision can be obtained with stationary power saws, most of the procedures can be completed satisfactorily using only basic hand power tools.

Building a boxed post There are a number of reasons to build a boxed post. The most common is to add a flamboyant touch to an elaborate period fence, but another reason is encase a pressure-treated post. Even posts of #1 grade southern yellow pine are prone to check and twist over time. Boxing the pressure-treated post in good, finish-grade pine, cedar, or

Boxing a post is an elegant way to dress up a fence.

redwood will produce a much more attractive post. A final reason to box a post is that it can be less expensive than using solid stock when a 4x4 is structurally adequate but the design calls for an 8x8. The drawing below shows how to build the boxed post.

Boxing a post with ordinary 2x dimension framing lumber defeats the purpose of encasement. Although it's possible to find spruce or fir that's near clear, the 2-in. dimension guarantees instability that will eventually cause the casing to cup or the corner joints to open. When framing-lumber casement boards are butted together at the post corners, the rounded edges of the two

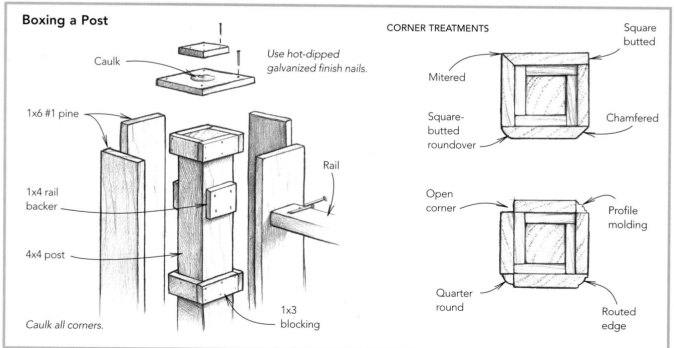

Boxing a Post

Caulk

Use hot-dipped galvanized finish nails.

1x6 #1 pine

1x4 rail backer

4x4 post

Rail

Caulk all corners.

1x3 blocking

CORNER TREATMENTS

Square butted

Mitered

Square-butted roundover

Chamfered

Open corner

Profile molding

Quarter round

Routed edge

1. Use a hard-board template to trace the outline of the finial on the post.

2. Cut the finial on the bandsaw.

3. Cut a decorative vein around the base of the finial by making opposing 45° cuts on the radial-arm saw.

4. The completed finial.

pieces will form a prominent and not very visually appealing groove. Mitered corner joints eliminate this problem, but given the instability of the lumber, any such joints will soon open. Also, mitered corner joints cannot be used with chamfered or beaded corners. The drawing on p. 174 shows some alternative corner treatments.

Veining a post Once the post is cased (or even if it's not), an option is to cut decorative veins and grooves along its length. The router is the tool of choice for adding these details. The large selection of cutting bits available allows you to fashion distinctive V-groove, round and flat-bottom, half-round, and profile-beaded narrow veins and grooves of various widths. (When thick stock is required as a base for deep veining, use stable, finish-grade 5/4 or 6/4 pine.)

To cut a vein or a groove in a post, use the router's parallel guide attachment. Adjust the guide so that the bit lines up with the groove location. Reset the guide to make a series of parallel grooves. If the groove does not run the entire length of the post, set stop blocks at its starting and ending points.

Decorative veins can also be cut across the grain of the post, as when detailing the base of a finial (see the photo essay on p. 175). For this cut, you can use a router, a radial-arm saw, or a circular saw.

Chamfering a post Veins and grooves are decorative details applied to the face of the work. A chamfer is a decorative detail applied to its corners. Unlike veins and grooves, chamfers almost never run the length of the piece. Indeed, it's the singular appearance of the stop that accounts for a good part of chamfering's visual impact. Other than this distinction, there's no real difference between the methods for cutting these two different details. Since the cutter is designed for edge work, chamfering bits are also available with a self-guiding roller bearing that does away with the need for a parallel guide.

Cutting rosettes Rosettes are a very striking post detail that can successfully evoke the look of a high-style Victorian fence on the cheap. This is another detail that can be cut with a router. To inlay a rosette on a post, locate the center of the rosette and drill a 5/16-in. hole for the router trammel guide pin. Choose the bit pattern (a corebox bit is a good one), attach the trammel guide to the router, and place the guide pin into the pivot hole drilled in the post. Change depth, pivot-arm length, or bits to shape the design.

Making a post cap Although you can buy premade post caps, the selection is pretty much limited to what will fit onto a 4x4 post, and the choice of styles is limited to a few albeit often quite handsome, basic variations on classical revival themes. Making your own caps can be a lot more fun.

Since 2x stock will be required to mill a post cap of any substance, only the best, firmest-grained knot-free wood will withstand the rigors of exposure without checking and warping. You can usually cut suitable blanks out of planks of #1 treated lumber. Unlike the face of the post, this is one place where appearance might reasonably be sacrificed to durability. Otherwise, if you opt for untreated finish-grade wood, seal all hidden surfaces with a brush-on preservative before assembly and caulk all joints. As long as you repaint it at regular intervals, the cap should last at least 50 years.

To make the cap, start with a square blank and cut the top bevel. Short bevel cuts can be made with a router fitted with a raised-panel cutting bit (used to shape cabinet-door panels). Longer bevels should be made on the table saw. Set the blade at the desired angle (from 10° to 30°) and slide the blank on edge through the blade, holding it against the saw fence. The cut will be safer and more accurate if you use a sliding fence or tenoning jig.

From here, you can add whatever level of ornamentation your heart desires. You could cut decorative molding on the underside of the cap on a router table, add crown molding or dentils between the post and the underside of the cap (as shown in the photo on the facing page), or top the cap with additional layers of cap stock and an

Add dentil molding under the post cap for an elegant detail.

ornate finial. In terms of function, a simple square post cap slightly larger than the cross section of the post is as effective as the most elaborate confections of the classical and Victorian styles, but it's not as much fun to build. Cap detailing is one place where you can play at woodworking.

Fancy pickets Depending on the intricacy of the cutouts, several different tools can be used to make fence pickets that do more than come directly to a point. A router can generally handle most patterns in stock up to ¾ in. thick. Thicker stock requires a bandsaw. The first step is to make a template for piloting the router or marking the cut lines for the bandsaw.

Begin by tracing the pattern onto the template (I use ¼-in. hardboard, as shown in the top photo on p. 166). Accuracy is of the utmost importance in template layout. Any error here will clone itself endlessly. Where the pattern calls for a round hole, make it perfectly round: Use a drill press, not a hand drill, scrollsaw, or jig saw. Use a scrollsaw or jigsaw for making interior cuts, and a bandsaw or jigsaw for exterior cuts.

To cut the picket with a router, screw or clamp the template to the stock and use a self-guiding bit. Set the bit depth so the roller bearing glides on the edge of the template. A carbide-tipped double-fluted straight-cut bit will stay sharper longer and make

cleaner cuts without overheating than a less expensive high-speed steel bit.

MAKING A DOWELED PICKET FENCE

Using dowels instead of flat or square pickets is an interesting variation on the standard picket fence. A doweled picket isn't that difficult to build; the key is to bore perfectly matched and centered holes in the top and bottom (and optional middle) stringers of the fence. For this you'll need a drill press (though you could get by with a hand drill mounted in a drill-press stand).

Set the depth stop of the drill so that the holes go all the way through the top stringer and one-third of the way into the bottom stringer. Clamp the stringers on top of each other, with additional side clamps to straighten out any bows and keep the sides flush if necessary. After drilling the holes, bevel the stringers so that they'll shed water away from the doweled infill.

Use ¾-in., 1-in., or 1⅛-in. hardwood or pressure-treated dowels. Sand the tops smooth, making a slight chamfer by rotating the edge against a stationary belt or disc sander.

Install side-frame members at the ends of each picket panel. (If you prefer, the top stringers can overlap the posts, with the bottom stringers set in notches in the post.) Assemble the panels in the shop, leaving out the last three or four dowels to provide room for screwing the side frame to the post. Glue the dowels to the bottom stringers with exterior-grade construction adhesive.

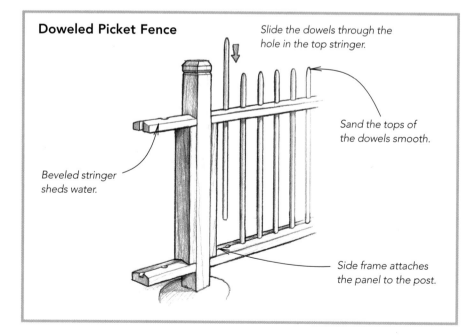

Doweled Picket Fence

Slide the dowels through the hole in the top stringer.

Sand the tops of the dowels smooth.

Beveled stringer sheds water.

Side frame attaches the panel to the post.

An option when building a doweled picket fence is to stagger the height of the dowels.

MAKING A LOUVER FENCE

Vertical and horizontal louver fences provide a good deal of privacy and environmental buffering without appreciably restricting airflow. Because an intermediate stringer cannot be used in louver fences, the unsupported louvers must be cut from stable, kiln-dried, high-grade wood (ideally, redwood or cedar) if they are to resist warping.

There are two ways to build a louver fence: The easiest is to use 1x spacers between the louvers; the more elegant approach is to house the ends of the louvers in dadoed slots cut in the stringers (vertical-style louvers) or posts (horizontal louvers), or in a vertical frame piece that attaches to the posts. These slots can be cut either on a radial-arm saw with a dado blade or with a router and a shop-built jig.

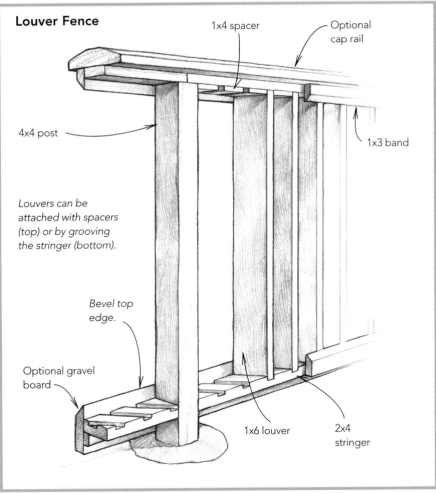

Louver Fence

1x4 spacer

Optional cap rail

4x4 post

1x3 band

Louvers can be attached with spacers (top) or by grooving the stringer (bottom).

Bevel top edge.

Optional gravel board

1x6 louver

2x4 stringer

Charles Miller

Charles Miller

Tharp-Hamilton Woodworking

seven

GATES

There is but one

right way to brace a gate,

and many wrong ones.

George A. Martin, *Fences, Gates and Bridges: A Practical Manual* (1887)

If you've built your fence right, it's apt to stay where you put it, unless something big and strong gives it a shove. But gates are perverse. In a fundamental sense, a stable gate is an impossibility: The laws of physics conspire against it. Quite simply, gates tend to sag. And when they do, they don't work right anymore: They pinch against the latchpost, the latch mistakes the strike, the very boards and battens that had once come together in close harmony seem to have acquired a repulsive charge.

THE PHYSICS OF A GATE

A gate will sag if the gatepost shifts or the gate frame is racked by the force of its own weight. Only an angled cross brace can confer the stiffness that empowers the gate to resist. Lacking such a brace, the rigidity is sustained solely by the strength of the corner joints and the tenacity of the fasteners that hold them. It's not enough. Over time, the tug of the gate will draw the nails and chafe the joints; the grip slips,

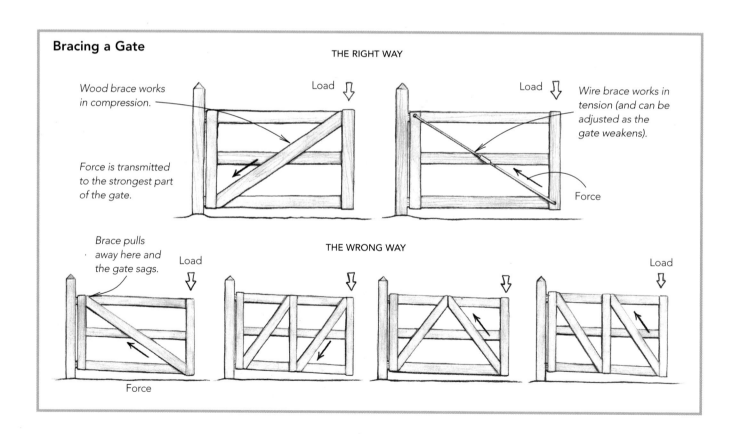

Bracing a Gate

THE RIGHT WAY

Wood brace works in compression.

Load

Load

Wire brace works in tension (and can be adjusted as the gate weakens).

Force is transmitted to the strongest part of the gate.

Force

THE WRONG WAY

Brace pulls away here and the gate sags.

Load

Load

Load

Load

Force

This farm gate is correctly braced and will resist sagging.

and the erstwhile rectangle slouches into a dissolute parallelogram.

Gates must not only be correctly braced but also be braced in the right direction. Judging by the plethora of gates whose brace runs from the top at the hingepost side to the bottom at the gatepost end, quite a few fence builders apparently were unfamiliar with physics. If the brace is installed in the wrong direction, downward load on the end of the gate will be transmitted along the diagonal brace from the bottom at the latch side up to the top at

the hinge side. The stile (the vertical frame member) at the hinge side of the gate will be pushed sideways, augmenting the already considerable upward force on the hingepost. It's little wonder then that pushing down on the end of the gate translates into a tendency for the bottom hinge to rip from the gatepost and for the post itself to rise up out of its hole.

When the brace runs from the bottom of the gate at the hingepost to the top at the latchpost end, the downward load is exerted against the

Although this walkway gate is attractive and well built, the angled brace is still installed in the wrong direction. (Photo by Bill Rooney.)

strongest and most stable point of the post. The fulcrum is too close to the handle of the lever to give enough purchase for a good lift. In fact, there's a tendency to push the hingepost downward. The gate might be torn from its hinges, but it won't sag, and the post, if well footed, will remain steadfast and true.

The correctly positioned wooden brace shown in the drawing on p. 184 works because it's in compression. For a gate so braced to sag, the brace would have to shorten, which, given the extreme strength of wood in compression, is impossible, at least under the conditions that exist within

a gate. A gate can also be braced in tension, that is, by pulling the opposing sides together. This can be accomplished by running a threaded steel rod or wire rope and a turnbuckle diagonally from latchpost to hingepost, exactly opposite to the correct direction of a wooden brace. Tightening the turnbuckle pulls the corners of the gate together. If well built and correctly hung, the gate will withstand downward loads without sagging.

A significant advantage of this kind of tension brace over a wooden compression brace is its light weight (the best gate is one that is lightest in proportion to the greatest strength). Another advantage is the ability to adjust the brace to take up slack. Wooden braces are generally acknowledged to harmonize better with a wooden gate than steel tension braces. However, in the case of a gate for a picket fence, a wooden brace could arguably disrupt the rhythmic pattern of the vertical pickets, whereas viewed through the pickets a tension brace would be almost invisible.

Two other factors collude against the stability of the gate. First, a gate under load will attempt to relieve itself of the strain. According to the laws of structural mechanics, since a vertical beam fixed at one or both ends cannot respond to a bending moment at its midspan by uplifting at its ends, it will deform, that is, twist and buckle sidewards along its length. Structurally speaking, a gate is a thin "beam," fixed at one end to the gatepost and at the other to the latch. Loading a gate is like pushing down on a sheet of paper held on edge.

Second, the hygroscopic properties of wood can cause a similarly unpleasant reaction. Wood doesn't shrink appreciably with the grain, but it does shrink acutely across it. Vertical boards screwed and glued to horizontal rails are stressed as they contract and expand with changes in the ambient humidity. In extreme cases, nails "pop" or boards split when their fasteners fail to give. The more usual response is for the gate to twist across the plane of its face. Using methods and materials that tend to reduce the inherent stresses on a gate will keep the damage at a tolerable level.

GATEPOSTS

The best-made and best-braced gate will come to naught if the gatepost to which it is hinged is not sufficiently strong and well anchored. Because they bear a significantly heavier sideward load than line posts, gateposts are often thicker and always more deeply set than ordinary fence posts. The requirement for greater cross-sectional area generally applies only to long and heavy farm and driveway gates. Standard line posts, as long as they are set at least 1 ft. deeper, can serve as gateposts for a normal-width walkway gate.

It may be that the heft of a gate mandates a beefed-up post, but the visual integrity of the fence is felt to be unacceptably compromised by the

Gateposts may need to be thicker than standard line posts to hold the extra load.

resulting "bump-out." A possible way out of this quandary is to increase the post width only in the dimension of the strain (for example, using a 4x6 instead of a 6x6). This would moderate the offending aspect while retaining more than enough of the structural strength of the greater post.

Gateposts are often taller than line posts as well (see the top photo on p. 142). The ostensible rationale may be aesthetic: Increasing the height of a gatepost emphasizes its importance in the scheme of the fence and helps the caller find the gate. Tall gateposts spanned by a horizontal rail are an archetype of the temple portal, an icon of the sacred threshold separating the inner and outer realms. The evocative and often-replicated image of the

Bernard Levine

Bernard Levine

The height of a gatepost is sometimes increased for structural purposes, as on this aging farm gate.

Shinto *torii* is a well-known example of this principle in action. The rose-clad, overarching trellis gateway is its homelier, more mundane application.

The height of a hingepost can also be increased for structural purposes. Wide farm gates are often reinforced with a cable and turnbuckle-type tension brace extending from the top at the latch side to an anchorage high up on the post. The steeper the angle of the diagonal (that is, the shorter its length) the greater the bracing effect. This kind of tension brace can also be made from wood, as shown in the photo above. Note that the hingepost stile must also be lengthened to support the brace. Regardless of the type of tension bracing used, the hingepost must resist the sideways bending force of the brace, either by added stiffness or by diagonal bracing running back onto the flanking fence panels.

ANCHORING THE GATE

Whatever the height or thickness of a gatepost, its solid anchorage is the highest priority. In unstable or wet soils or soils subject to frost action, increasing posthole depth may not be enough to prevent a heavy gate from kicking the post out at its bottom. Incorporating a strain plug into the post footing can provide the extra reinforcement needed to keep the post from moving and the gate from sagging (see the drawing at left below).

To ensure solid anchorage, begin by digging the postholes at least 3-ft. deep (in relatively frost-free zones; otherwise, dig deeper as frost penetration and post length require). The hole should be about four times as wide across as the post is thick (e.g., 16 in. for a 4x4, 24 in. for a 6x6). Nail a pair of treated 2x4 or 2x6 crossbars as long as the hole

is wide to opposite sides at the base of the post, and then drive a pair of rebar pins from one side through to the other to prevent the crossbars from twisting on the nails. Tamp the bottom of the hole, add 6 in. of crushed chestnut stone, set the posts so the crossbars line up with the "drag" line of the gate (the same as the fence line), and continue tamping in alternating layers of earth/gravel and crushed stone, stopping about 12 in. below grade.

Now form and pour the strain plug the width of the post, the length of the hole, and 6 in. to 8 in. deep. Before mixing the concrete, drill the post and

drive two parallel ½-in. rebars cut about 4 in. shorter than the form through it. After the concrete has set and the forms have been removed, spread another layer of stone, and then backfill to the surface with heavy clay earth. The crushed stone provides drainage, and the crossbars and concrete strain plug help keep the post from creeping sideways through the soil under the pressure of the gate.

In rain-soaked or frost-prone soils, you might want to consider pouring a monolithic concrete threshold post anchor (see the drawing at right below). Locked forever in a solid block of

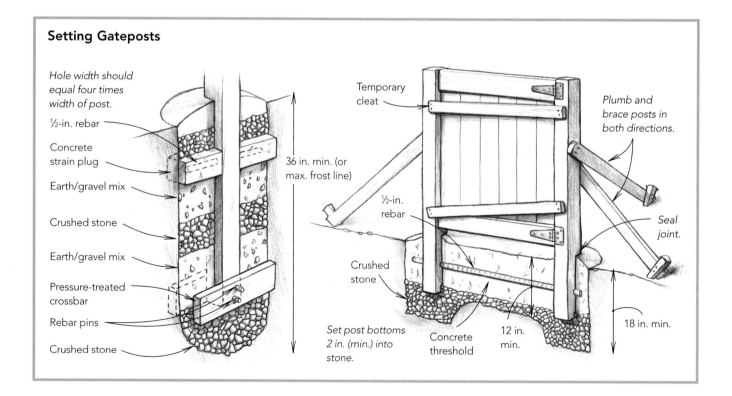

Setting Gateposts

Hole width should equal four times width of post.

½-in. rebar

Concrete strain plug

Earth/gravel mix

Crushed stone

Earth/gravel mix

Pressure-treated crossbar

Rebar pins

Crushed stone

36 in. min. (or max. frost line)

Temporary cleat

Plumb and brace posts in both directions.

½-in. rebar

Crushed stone

Seal joint.

Set post bottoms 2 in. (min.) into stone.

Concrete threshold

12 in. min.

18 in. min.

concrete, the gateposts will always stay square to the gate. The gate (if poorly built) may sag of its own accord, binding at the hinges or latch strike, but the gateposts will be absolved of blame. If not footed below the frost line and well drained, the entire block, posts and all, can rise, fall, or tilt with soil movement or frost heave. This could play havoc with the visual, if not the structural, integrity of the adjoining fence sections, but the gate will always remain true to its posts.

To install a concrete threshold anchor, begin by excavating a trench three or four times as wide as the gateposts along the entire width of the gate and extending at least twice the post width beyond them. Dig at least 18 in. deep at the post centers and no less than 12 in. deep at the center of the gate. Locate, plumb, and brace the posts on the tamped crushed-stone base. Secure the gate on its hinges to the hingepost, using supporting blocks as needed. Position the gate where it will be when closed and tack it in place with temporary cleats between the posts.

Drill through the posts for a ½-in. rebar dowel (if the sides of the hole won't allow you to insert a single length, overlap two shorter dowels and tie them together). Tie the rebar to parallel dowels. Add a second course of rebar when the concrete is more than 12 in. deep. Rebar won't actually prevent the block from cracking but it will hold the pieces together afterwards. More important, the bar maintains the structural connection between the post and anchor even if frost manages to split off a piece adjacent to the post. Trowel off the fresh concrete to form a raised threshold between the posts and to slope away around the rest of them. Leave the braces in place for at least three days, and, after a week, seal the joint at the posts with silicone caulk or hot asphalt.

THE WELL-BUILT GATE

After properly anchored posts, a strong frame is the most critical component of the well-built gate. There are two basic types of gate frames: the Z-frame and the perimeter, or "box," frame. Each has its strong points and drawbacks.

This box-frame gate with interesting infill lends an air of elegance to a simple fence.

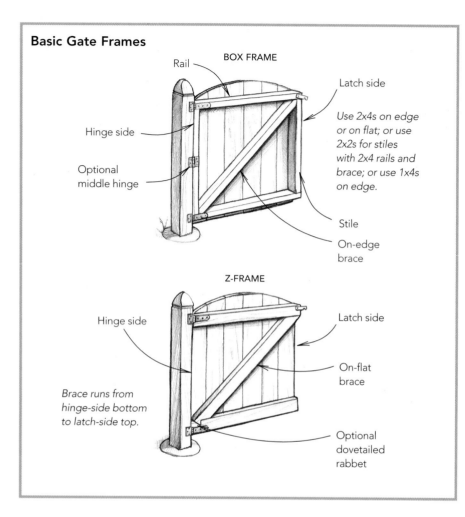

Basic Gate Frames

BOX FRAME

Rail

Latch side

Hinge side

Use 2x4s on edge or on flat; or use 2x2s for stiles with 2x4 rails and brace; or use 1x4s on edge.

Optional middle hinge

Stile

On-edge brace

Z-FRAME

Hinge side

Latch side

On-flat brace

Brace runs from hinge-side bottom to latch-side top.

Optional dovetailed rabbet

The Z-frame isn't as rugged as the box frame and should be used only for small (3-ft.-wide or less), lightweight gates. Z-frames can have an informal, unpretentious, country-cottage look, whereas box frames often feel much more deliberately "architectural," elegant, and sophisticated. Box frames must be used with inset infill (that is, with the gate infill centered into the frame attached to cleats rather than overlapping it).

As you ponder whether to build a box-frame or Z-frame gate, bear in mind that even with an inset infill, a 2x4 on-edge frame needs more clearance to close without binding on the latchpost than a thinner frame would. Experienced door-hangers are familiar with the problem of swing-arc clearance, which is why the leading edge of an entrance door is given a 5° bevel.

Constructing the frame on flat can beneficially reduce the thickness of a gate's edge profile as well as solve any clearance problem. Although it might also seem to follow that, as with fence stringers, an on-flat frame will make a stronger gate, the gain in stiffness is offset by a loss of strength at the butted corner joints, where it's much harder to make a solid mechanical connection than with on-edge joints. Toenailing risks splitting out the face and ends, especially with 1x stock. Furthermore, for gates up to 8 ft. wide, the risk of an on-edge frame sagging in midspan is a lot lower than the risk of an on-flat frame sagging at its corner joints. And since, unlike stringers, the ends of a gate frame aren't rigidly fixed to posts, an edgewise frame can help keep a wide gate from bellying, which is a particular problem when the gate is in motion.

Fortunately, there are ways to stiffen the resolve of on-flat corner joints so that you can build a gate that doesn't look clunky but still holds its shape. In fact, even fairly heavy gates can be framed with 1x4 and 1x6 boards on flat (although they do tend to bow considerably at right angles to their drag line). For the average walkway gate, a combination of 2x4 top and bottom rails and 2x2 stiles (vertical rails) or even 1x4s throughout will be more than rugged enough. Sufficiently rigid bracing is more important in this case than the depth of the framework. The addition of stiles to what would

The horizontal midbraces of these paired Z-frame gates provide support as well as visual continuity with the fence.

otherwise be a standard Z-frame endows it with the strength of the box frame (although the inherently weaker joint condition still remains) while keeping a lower profile.

GATE JOINERY

Butt joints are the most commonly used gate corner joints because they're the easiest to make, but they don't always provide a solid mechanical connection. You can use steel angle braces to reinforce the joints, but only at the expense of appearance. Their aesthetic assault can be mitigated by routing them to install flush with the face of the boards. Another alternative is to screw and glue triangular ¼-in. plywood gussets over the braces. Since you can't hide a gusset brace, consider transforming it into a creative design element: There's no law that says a

gusset has to have a straight hypotenuse. Let it curve and wander, make it big enough to sport decorative cutouts.

Half-lap, rabbet, dovetail, and mortise-and-tenon joints do require more skill and time to make than butt joints, but the effort is justified by the stronger connections that result. Lap joints are a good choice for on-flat frames. Rabbeted joints are excellent for on-edge ones. A dovetail notch is particularly effective for making strong on-flat connections between the cross brace and rails.

Whatever joints you choose, always fit stiles (verticals) in between rails (horizontals). This helps protect the joint from water infiltration. Also, the physics of the normal gate operates to ensure that the stiles carry relatively little of the load. When the gate is properly braced, the triangle under the

The first step in building a strong gate is to use good-quality wood. This is one place where it doesn't pay to scrimp on materials. If your fence boards were merchantable or construction-grade redwood, buy clear all-heart for the gate. Use only structural-grade dimension lumber for frame members. (Such lumber is specially graded for use with greater-than-normal building stresses.) Whatever the species, selecting premium kiln-dried boards with tight, straight grain and few or no knots will reduce the chance of their warping.

upper rail transmits no real load to the lower rail other than its own weight, which is about half the dead load of the total fence. As a compression brace in a small gate, a stile is largely redundant. A better rationale for including it is to provide more solid attachment possibilities for latches and hinges than the infill would alone. The stile also allows an on-flat 1x gate to close flush against the stop, without the gap that would otherwise exist between the rails and the infill boards.

DESIGN CONSIDERATIONS

Because of its visual and symbolic importance within the landscape of the fence, the architectural aspects of a gate's infill design are even more deserving of the sort of detailed

attention already lavished on the fence infill. As explained in Chapter 2, you can build your gate so that it blends in with the flow of the fence or deliberately contrasts with it. The gallery of gates shown on pp. 195-197 should give you plenty of design ideas.

On a more practical level, the slope of the ground can determine which will be the gatepost and which the latchpost and the direction of the gate's swing, as well as how to treat any infill projection at the bottom of the gate. Regardless of ground slope, the gate rails must always be installed level. Nevertheless, convenience and handedness or site limitations may require the gate to be hung on the uphill side of the slope. In this case, either cut back the slope to provide enough room for the swing arc

Setting the gate back from the line of the fence adds visual interest to an otherwise plain fence.

The visual simplicity of this gate remains uncluttered by diagonal bracing because the gate panels are carried by latchposts that bear directly onto the driveway.

or close the gap by leveling the walkway at the threshold and comfortably beyond the gate swing in both directions, and resign yourself to living with a gate that will only open to 90°. In most cases, a 90° arc isn't a particularly irksome limitation. In fact, the swing of any gate hinged between posts is limited perforce to this arc. Only gates hinged to the face of a post can open to 180°.

It's also a good idea to allow a few extra inches of clearance under the gate at a paved walkway or driveway so that you can sweep or hose it down without opening the gate. The extra height also leaves room for winter ice buildup. In any case, the gate should clear the

walkway by at least an inch as a hedge against binding if it ever sags.

Two final considerations on the subject of gate swing are handedness and direction of swing: Are you right-handed or left-handed? People carry things in their off hand so as to reserve the favored one for opening the latch. By custom, gates almost always open into the enclosed area. From the caller's point of view, this is only natural. Forced to pull a gate outward, callers must obstruct their own entry, which is hardly inviting. Hence the admonition of 19th-century architect Henry Cleaveland: "Unless you wish to invoke curses on your head, both loud and deep, don't let your gates swing outward!"

For convenience, a gate should open into the enclosed area.

GATE HARDWARE

When it comes to hardware, good looks and a strong body are important. Hinges and latches are like the accessories that complete and complement a stylish outfit. But hardware is also a critical structural component of the gate system. The scale of the hardware must be suited not only to the gate's appearance but also to its mechanics.

Hinges and latches are typically available in a number of matching finishes besides the standard plated steel, which you should avoid in any case since it will soon rust. Besides hot-dipped galvanized, which is a good choice for the average gate, finishes include black, bronzed, or antique verdigris (which looks like corroded copper). Ornamental finishes are typically available in matte (smooth) or antiqued (pocked, pebbly grain) styles, variously figured to mimic hand forging. Whatever their merits, the faux-forged look is not often convincing and, to my mind, not worth the exorbitant surcharge. If I wanted hardware with a hand-forged look, I'd have a blacksmith forge me some. The difference in cost is not as much as you'd imagine, and the results are worth it. Sometimes, custom forging is the only practical alternative, as when you need a set of very stout and extra-long strap hinges to carry a similarly stout gate. Hand-forged or machine-made, good hardware is always expensive and the flimsy economy-grade stuff is a poor bargain.

Using butt hinges mortised into the gate frame and post is a common way to hang a gate.

The wide bearing area of this traditional strap hinge helps to stiffen the gate.

HINGES

There are three basic kinds of hinges suitable for gates: butt, strap, and lag. Butt hinges are the familiar flat, rectangular plates (leaves) linked together by a hinge pin inserted through the hinge barrel. They can be either face-mounted or mortised door-style into the edge of the gate stile and the side of the post. In either position, butt hinges will swing only one way. Since they make gate installation more convenient, loose-pin butts (the hinge pin is removable) are preferable to fixed. Be sure to install them right-side up so that the hinge pin doesn't fall out. Most so-called "utility" butts are fixed pin only.

Butt hinges are not a good choice for gates much over 3 ft. wide. Weight isn't so much the problem (there are butt hinges that can handle 100-lb. doors) as the compounding leverage effect of increasing gate width on the narrow hinge leaves, which tends to loosen their screws.

Strap hinges are a much better choice for all kinds of gates. Their long plates distribute the load over a much wider area with much less angular leverage than butt hinges. However advantageous the long strap may be when carrying the gate rail, it's a problem when it's longer than the gatepost is wide. Although you can bend the hinge to wrap around the corner, a T-strap hinge is the normal, and more elegant solution. The long strap of this combination hinge bears the heavy load of the gate proper, while its rectangular leaf fits the post. All strap hinges have fixed pins. A wide choice of sizes, styles, and finishes are available to suit just about any type of gate.

Pintle-type lag hinges are the no-nonsense workhorses of the gate-hinge community; they are an ideal choice for use with round posts. The stout (⅜-in. and up) L-shaped male half, or pintle, screws into a pilot hole drilled in the gatepost. The female eye half is either through-bolted to the face of the gate or threaded into its stile. Either way, the eye is simply dropped down onto the projecting pintle. Because there are no small screws and hinge barrels to come loose, lag hinges will carry very heavy gates without sagging. Their sheer heft, however, requires an equally strong gate frame and post anchorage. A 1x stile is too thin for edge attachment—use a through-bolted strap-style lag hinge instead (as shown in the bottom photo at right). The same goes for a 4x4 post. Setting a pintle-type lag hinge in the gate stile allows the gate to swing in either direction in almost a full circle. The swing arc of a face-mounted lag is limited to 180°.

Pintle-type lag hinges can carry very heavy gates without sagging.

Here, a through-bolted strap hinge is carried on a lag-bolted pintle.

LATCHES

There are three basic kinds of latches, each differentiated by how they mount on the gate: face-mounted, through-mounted, and bailed. There's probably a technical distinction between latches, catches, and bolts, too, but it's not worth belaboring, so long as they know what you're asking for down at the hardware store. Most people use the terms interchangeably in any case.

When choosing a gate latch, besides suitable style and visual appeal, consider mounting requirements (how much of the hardware will be concealed

A thumb latch is the most common kind of through latch.

An elegant wooden face-mounted latch.

Although more commonly fashioned from wire or bent-wood hoops, a bail latch can also be as simple as a couple of loops of rope.

or revealed), convenience of operation, security requirements, and the thickness of the gate stock. Some latches can only be face-mounted, while others will only mount on the back of the gate. Thumb latches and other sorts of through latches will not work in thick panels (such as box frames), unless the thumb lever is custom-lengthened.

Face-mounted latch styles run the gamut from hooks to slide bolts and even include door-style knob-handled locksets. They all mount on either face of the gate and post. String-operated pivots (the proverbial "latch-string") and thumb latches are two of the most traditional types of through latches, which operate by opening a catch on one side of the gate from the other by means of a lever inserted in a hole bored through the gate. Bails are wire or bent-wood hoops similar to the handle (more properly known as a "bail") on a paint pail. The bails pivot on a short post or gate to loop over an extension of the gate stile or the top of the post, respectively.

STOPS

As with any door, the stop, typically a vertical piece of wood attached to the latchpost or gate face, reduces stress on hinges by stopping the gate before it can overswing its normal operational arc. Although you can use the same sort of wood for the stop as that of the gate, since the stop is subjected to considerable abuse from repeated collisions with the gate stile, a hardwood strip might be a better choice.

FASTENERS FOR GATE HARDWARE

The screws prepackaged with most gate hardware are seldom long enough or stout enough to withstand the stresses of normal gate operation. Also, the chromed or electroplated finish of standard screws will soon rust. Since the zinc coating would gum up the threads, hot-dipped galvanized screws aren't available, at least in any size smaller than the thinnest lag screw (¼ in.). Stainless-steel screws are the best defense against rust. Although brass is equally rust-proof and looks exceptionally handsome in redwood and red cedar to boot, the metal is much softer: Overturning the screw even slightly can snap the head.

Use the longest screws you can. At the very least, they should be long enough to penetrate at least halfway into the wood or, better yet, within a ¼ in. of coming out the other side.

Gate Stops

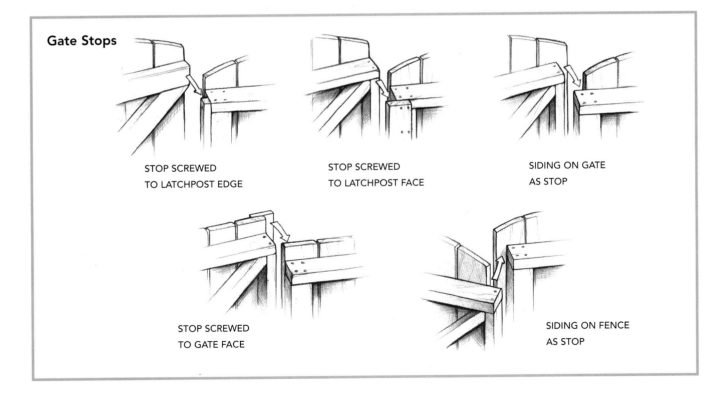

STOP SCREWED
TO LATCHPOST EDGE

STOP SCREWED
TO LATCHPOST FACE

SIDING ON GATE
AS STOP

STOP SCREWED
TO GATE FACE

SIDING ON FENCE
AS STOP

The drawing on p. 203 shows five possible options for mounting a gate stop. The stop screwed to the inside edge of the latchpost (the first option shown) is much stronger than any stops made by extending or battening the infill or post face.

HARDWARE FOR HOLDING GATES OPEN AND SHUT

Sometimes it's desirable to install hardware that will hold a gate open as well as keep it shut. This hardware not only makes repeat trips through the gateway convenient but can also prevent the gate from being torn off its hinges or from splitting the post or stile when the gate is violently thrown open or slammed shut by a gust of wind.

The simplest type of hardware for this purpose is a hook and eye, which is perfectly adequate as a gate saver. It's certainly a lot better than the traditional "stumble-peg," a length of pipe or a wooden stake driven into the ground to check the gate's outward swing. Like an interior door stop in the middle of the hallway, this eponymous device can be more of a hazard than a help. The same is true for the trusty old brick or boulder.

Cane bolts are a requirement with any double gate, both for security and for ease of operation (see the drawing at left). Finally, a spring closure or spring-loaded, self-closing hinges will put an end to the annoyance of gates that remain open because someone forgets to close them.

BUILDING A GATE

Now that I've explained what constitutes a well-built gate, it's time to put it all together. For the most part, building a gate is not that much different from building a section of infill (see Chapter 6). I've broken the process down into 10 easy steps.

Before you begin, consider whether it is better to build the gate in the shop, with the benefit of a flat work surface-cum-set-up jig, or in place between the actual posts it must fit. (These are the same options you face when building a section of infill.) The first method

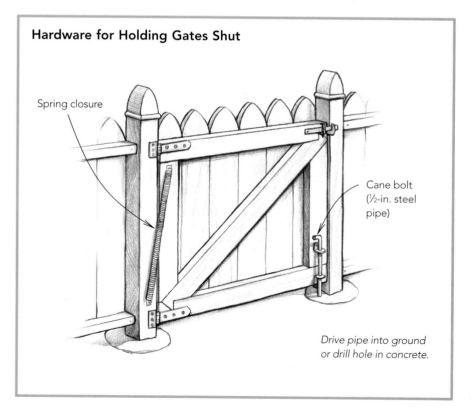

Hardware for Holding Gates Shut

Spring closure

Cane bolt (½-in. steel pipe)

Drive pipe into ground or drill hole in concrete.

requires careful measurements, plumb posts, and the skill to make and keep the gate square as you put it together. The second is a good choice when the posts are off plumb and the gate width varies from top to bottom so that the finished gate must be a trapezoid with level rails and skewed stiles. Building in place may be easier for the novice, but it will certainly take more time.

THE 10 STEPS

1. Measure the width of the gate opening at the top and bottom rail levels. If the measurements aren't the same, check the posts for plumb. Replumb the post(s) if possible or resign yourself to building the gate to fit the opening. Deduct the required swing-arc and hardware clearances from the overall width to find the rail lengths. Standard gate clearance is figured at ¾ in., which allows ½ in. at the latch side for the swing and ¼ in. at the hinge side to keep the stile from pinching against the post.

To figure the hardware clearance, hold the hinges (and latch) against scraps to represent the gate stile and the latchpost and gatepost, and measure the thickness of the hinge barrels, folded leaves, or latch strikes. Hardware clearance generally isn't a significant problem with round posts since the curve of the post provides built-in clearance.

2. Cut and prepare the frame components. To keep water from damaging the joints, always overlay the rails on top of the stiles. Coat all joints

and cut ends that butt each other with preservative water repellent or paint before assembling them. Paint any metal too, if needed. (Hold off painting the rest of the gate for about a week after it's hung to allow any residual moisture to escape.)

3. Square and assemble the gate. You can use a carpenter's framing square to true the assembled but unbraced frame or the old layout trick of measuring the diagonals: If they're equal the rectangle is square on all sides. However, for gates much over 4 ft. in width and height, any one corner that checks out by the square won't necessarily square up with the others, which is why I prefer to build my gates in a known squared jig on the worktable (this is the same jig I use

---

BUILDING A GATE IN PLACE

To build a gate in place, cut the top and bottom gate rails to fit snugly between the posts. Level the rails and toenail them in place through the ends (let the nail heads stick out). Attach the stiles, using opposing cedar-shingle spacers to hold the necessary clearance at the posts. Cut and install the diagonal brace.

Face-mounted butt hinges that attach to the stile but not the rail can be installed at this time. Otherwise, nail (or screw) on the infill and hold the gate in place with a pair of temporary cleats tacked across the frame and posts. Pull the toenails and cut off the projecting rail haunches with a handsaw. Attach strap hinges to the back of the rails (or face-mounted on infill boards directly over them), check the gate swing, and attach the latch.

for the infill panels; see the photos on p. 168). The jig cleats hold the gate rails and stiles to the perfectly straight lines necessary to guarantee square corners.

Remove the squared (and temporarily braced) frame carefully from the jig, lay it on top of the diagonal brace stock, and scribe the cuts to fit. Return the frame to the jig, remove the temporary braces, and attach the permanent one.

4. Attach the infill. Check the orientation of the diagonal brace: It's upside down unless it runs from the hinge-side bottom to the latch-side top. As you do, remember to orient the gate properly; in most cases, you'll be working back side up. It's easier than you think to put the infill on the wrong side of the frame.

Leave the frame in the jig so that the infill will align itself square to the frame. Otherwise, recheck the frame for square and be sure that the infill runs parallel with it. For box frames, use two screws in each board at each rail. Drive fasteners every 6 in. to 8 in. apart along the edges over the stiles. Spread a bead of clear (or appropriately colored), paintable silicone caulk at the joint between the backs of the infill boards and the top edges of the gate rails.

For best appearance, install solid infill boards so that they end in equal widths at both sides of the frame. Let the tops and bottoms of the boards overrun the rails and trim them off after the panel is fully assembled, cutting from the back side with a circular saw.

If the hinges will be face-mounted on the infill where it does not directly overlay the frame, add solid blocking beneath them.

5. Mount the hinges on the gate first. Drill pilot holes for the screws. Use at least three hinges on gates 6 ft. and taller and secure them with heavy screws or bolts. Always attach hinges directly to the frame, never to the infill alone. For heavy gates, use oversized strap hinges and bolt them through the frame with carriage bolts, not lags or screws. For best appearance, counter-sink a hole on the nut side of the bolt so that the end of the bolt and its nut and washer lie beneath the face of the rail.

6. Check the fit. Prop the gate in its opening on blocks, with the hinges resting open against the post. Wedge the gate tight to the posts with wood-shingle shims, adjusting for proper clearance. Remove and trim for better fit if necessary. Otherwise, tack cleats across the gate to the posts on the side opposite the swing.

7. Hang the gate. When fastening the hinges to the hingepost, set only one screw in each hinge and check the gate for swing and alignment before driving the rest. Otherwise, each time a screw is reinstalled in a "used" hole, it loses some more of its grip. Install the top and bottom hinges before the middle hinge. Check the gate for ease of swing. If its back edge tends to bind against the hingepost, or if the gate resists effortless closure even though it doesn't actually bind against the hingepost, try shimming the hinge plates outward or springing the hinge knuckles. Binding is often a problem with gates hung from a round post.

8. Install the stop on the latchpost. Close the gate and set the latch side flush with the front face of the latchpost. Mark where the back of the gate frame meets the post and screw the stop to the line.

9. Add any supplemental tension braces. Complete threaded-rod tension-brace kits can be purchased at the hardware store. Since the braces are installed on the face of the gate frame, the rod ends will be flattened and drilled for screws. Any threaded tension rod that will be used as a pivoting supplemental brace must have screw-eye fittings installed. These are

usually custom-made by welding a length of rod to the appropriate end fittings (see the drawing on p. 225). Wire rope and clamp braces are easier to make but not as neat looking.

10. Install the latch, following the instructions on the latch packet. If you can afford the time, it's not a bad idea to wait a week or so for the new hinges to wear in a bit. Then any small sag will not require you to relocate the latch piece to keep it properly aligned.

The author's gate installed, complete with carved pineapple finials atop the gateposts as a symbol of "welcome." An unobtrusive threaded tension rod is yet to be installed.

eight

MAINTENANCE & REPAIR

…The gaps I mean,

No one has seen them made or heard them made,

But at spring mending-time we find them there.

I let my neighbor know beyond the hill;

And on a day we meet to walk the line

And set the wall between us once again.

Robert Frost, *"Mending Wall" (1914)*

This chapter is for those of you who are blessed with a fence that either through malfeasance in design or construction or just the general debilitation of age is in bad shape. Fence years are like dog years: Any fence more than 10 years old (even one built with pressure-treated wood) will have problems that must be fixed to arrest further decline.

Fence problems that are ignored don't go away, they get worse. Deterioration begins slowly at first, with a post that eases away from the wind in unstable soil, rails that sneak out of their notches, boards that melt into the ground. Rigid connections grow slack, the orderly fence line skews into irrational chords, and, like the increasing dishevelment of the

drunkard's progress, the fence slips and slides down the parabola of gravity until it reaches that moment of critical mass, that final confrontation with unyielding entropy, and it cascades into collapse.

Triage for the mortally ill fence comes down to the question of whether or not to pull the plug. Contractors use a rule of thumb to answer the question "Can this fence be saved?" If the repair cost is greater than 40% of the replacement cost, it doesn't pay to fix it.

ROUTINE MAINTENANCE

Fences, buildings, and automobiles are alike in that routine maintenance and preventative care are better than heroic rescue measures engendered by neglect. Regular tune-ups and oil changes are less trouble than rebuilding the engine. At a cost of $10 to $20 a foot, the investment in your hand-made front-yard fence justifies prompt attention to maintenance and repair of problems when they first appear.

Preventative care begins with an annual inspection. In the spring, after the frost has gone and the ground is still soft but not too muddy to dig in, walk the fence line and do the following four-point inspection.

CHECK THE POSTS

If the posts are made from untreated wood, you need to check for rot. For posts footed in concrete, probe at the base of the post with an ice pick or the

Some fences are so far gone that it makes more sense to replace them than to repair them.

This cracked post cap is badly in need of repair to prevent further water infiltration and splitting.

point of a knife. In soil, expose the first 6 in. or so of buried post. If the wood is spongy, it's rotten, and if the sponginess extends much more than ½ in. on all sides into the core, you've got a post to replace. If the rot is only superficial, however, scrape or chisel it off to expose sound wood, coat it with preservative, and check again next year.

Check the seal between the concrete and the post, rip out any shrunken or cracked caulk, and recaulk as needed. Check for loose or cracked footings that will have to be dug out and replaced.

Inspect the tops of the posts as well as their bases. Uncapped or missing or damaged post caps allow water to get into the post and rot it from the top down. Check the notches in the sides of the posts where the stringers are attached. These are also havens for rot, especially in damp climates, where they can acquire a mossy overcoat.

Next, check the posts for plumb. Even if a post is still vertical, push on it to see if and how much you can wiggle it, and if the movement is worse than last year. The post may need bracing or a footing repair to steady it. Look for soil settlement or softness around its base and add tamped fill and regrade.

CHECK THE STRINGERS AND INFILL
Look for loose joints between the posts and the stringers. Although straightening a leaning post will solve most problems, cracked, sagging, or badly warped stringers or rails must still be

replaced. Look for infill panels that might need to be braced or replaced. Note loose, warped, cracked, or missing boards and "popped" nail heads. Use this opportunity to renail loose boards. Unless you use thicker or longer nails, don't reuse the same nail holes. It's useless to hammer popped nails back down—they've already failed once and won't do any better with a second chance.

Moss growing on the bottom of the infill boards is a sign of encroaching decay. The backsplash from rain, moisture from snow piles, and overreaching vegetation are also threats. Look for and remove any dirt that has built up in the clearance zone. Remove the tendrils of vines that will otherwise

work their way between the boards and pry them apart. Enforce a regime of weekly weed-mowing or blaze an herbicidal no-man's land along the fence line. Look for evidence of wood-destroying insect infestation: the telltale sawdust castings, boreholes, and adobe-like termite tubes. Small-scale chemical warfare will work on carpenter ants, but termite extermination may require hired mercenaries.

CHECK THE GATES

You're probably already well acquainted with the symptoms of any problems your gate may suffer through daily use. The gate may sag or bind because of loose hinges, a wobbly post, or a frame that's pulled out of square (see the photo on p. 210). What's needed at this point is a thorough investigation to determine the under-lying causes, a plan for the repairs, and the resolve to carry them out.

CHECK THE FINISH

Loose or missing paint allows water to permeate the wood and foster rot. Don't wait until the fence paint has peeled and flaked off the entire fence before repainting; instead, spot-paint bare wood as necessary. Check joints and caulking for signs of failure, and recaulk as needed. Pay particular attention to the many joints of oversized ornamental post caps. Look for new checks in the exposed face of the cap and along the edge grain. Coat with preservative, caulk, and repaint.

As part of your annual fence inspection, check for sagging stringers.

Infill boards that extend all the way to ground or street level are particularly prone to paint failure.

Remove all the flaking paint before repainting.

Check the finish of the hardware. Remove any flaking paint or rust scale with a wire brush or chemical rust remover and reapply a new rust-resistant finish. At this time, oil all hinges and latches.

If repainting is necessary, first remove all the old blistered, flaking, or peeling paint. Small areas can be scraped by hand, but a rented pressure washer or heat gun is faster and easier. Always scrape in the direction of the grain, never across it. There's no need to remove paint that's firmly adhered, but the patch should be sanded so its edges will blend in with the new. When

using a heat gun wear a dust mask and gloves. Direct the heat onto the paint until it blisters (usually 5 to 10 seconds), and scrape off the loosened paint with a stiff-bladed putty knife, advancing the heat gun forward as you scrape. Wash down the fence with water and a mild detergent, rinse, and dry well before painting (for more tips on painting a fence, see pp. 225-227).

CAUTION: LEAD PAINT

Old fences, or more to the point, old paints, contain lead, which wasn't removed from paint formulations until about 1974. Exposure to dust and chips of lead paint or fumes from heat removal is extremely hazardous, and its introduction into the environment is irresponsible. If you are removing old paint from an old fence, get professional advice on the proper procedures for removing lead paint. State or federal environmental officials can assist you as well.

If the fence has a lot of paint buildup, there's a good chance it contains lead.

POST REPAIRS

Fence problems are most often post problems. Posts are the first thing to go—either by decay (until the advent of pressure-treated wood) or by being thrown out of alignment because of soil and wind conditions, structural failure under load, or invading tree roots. Misaligned posts loosen stringers and infill, stressing the entire frame system. Rot saps structural strength. The injured section becomes a dead weight pulling on the panels at each side, adding to the load on their posts, too. In places like California, you don't even have to do any thing wrong for your fence to fall out of line: The native "terra infirma" does it for you.

STEADYING FENCE POSTS

If the post is sound, fence-post wobblies are invariably caused by a weakened or improper foundation. In extreme cases, you'll have to dig out the old foundation and renovate the posthole by adding absent crushed stone or gravel backfill, or replace the old foundation with a stronger, more stable one. For gate and terminal posts, this usually involves adding some kind of strain plug (see pp. 190-191).

A leaning post is a sign that whatever footing it had, it wasn't good enough or done right in the first place. Don't replace an old footing with the same thing that failed once already. Complete foundation renovation is a measure of last resort. There are a

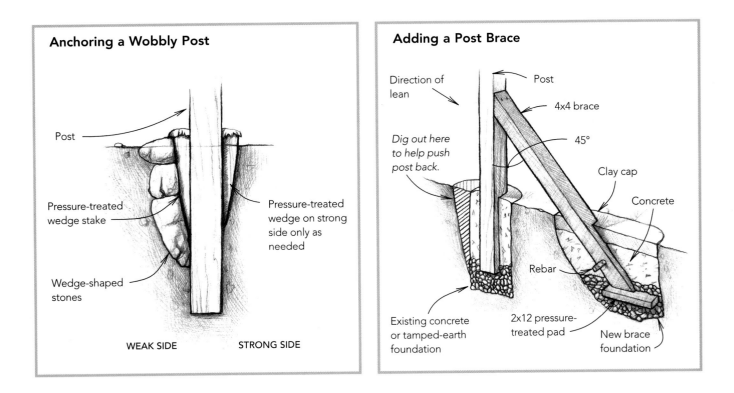

Anchoring a Wobbly Post

Post

Pressure-treated
wedge stake

Wedge-shaped
stones

Pressure-treated
wedge on strong
side only as
needed

WEAK SIDE   **STRONG SIDE**

Adding a Post Brace

Direction of
lean

*Dig out here
to help push
post back.*

Post

4x4 brace

45°

Clay cap

Concrete

Rebar

Existing concrete
or tamped-earth
foundation

2x12 pressure-
treated pad

New brace
foundation

couple of intermediate options to hold
the fence for another year.

Sometimes, particularly with fence
posts that are spiked into the soil rather
than planted in postholes, replumbing
posts heaved off-kilter by frost is a
seasonal activity, just another spring
chore that you can't do much else about.
Pushing or pulling posts back into
plumb by hand and tamping around
them will get you by until next spring.

Once the post is plumb, anchor it
by pounding wedge-shaped stones into
the soft earth along the weak side and,
to lock it in place, pound a treated-
wood wedge between the stones and the
post (see the drawing at left above).

Wedge the other faces of the post to
keep the pressure on the primary wedge.

If appearance is not of primary
importance, another quick fix for a
wobbly post is to drive a steel pipe into
the ground and brace the post to it with
a length of pipe hanger's strapping or a
wire rope.

A more permanent fix is to add a
post brace (see the drawing at right
above). This method is particularly
appropriate where a tall, solid-infill
fence fails to stand up to the press of
the prevailing wind. Fashion the brace
from the same stock (or the same cross-
sectional dimension, when adding
pressure-treated to untreated wood) as

the post. Dig an elliptical (egg-shaped) hole perpendicular to the face of the post on the side to be braced. For best results, make its bottom as deep as the fence-post hole. The center of the brace hole should be as far from the fence as the distance between the point where the brace attaches to the post and the bottom of the posthole.

Cut the 45° angle brace to fit, and then plumb the post. Digging away the earth on the opposite side of the post makes it easier to push it back into line. When the post is plumb and braced, form and pour a concrete strain plug (at least three times the post width and the full width of the hole) around the brace post to within 4 in. of grade. (Add rebar to fix the brace to its anchor.) After the concrete sets, cap off the hole with tamped clay soil.

REPAIRING SPLIT POSTS

Through-mortised posts of the kind used with split or peeled-pole rail fences are subject to splitting out when subjected to a strong lateral force or when severe checking caused by improper seasoning weakens the grain. Nails and bolts alone lack the power to draw the sundered post together or to keep it from spreading further apart.

To repair the split, wrap a chain as high up on the post as you can around

Repairing a Split Post

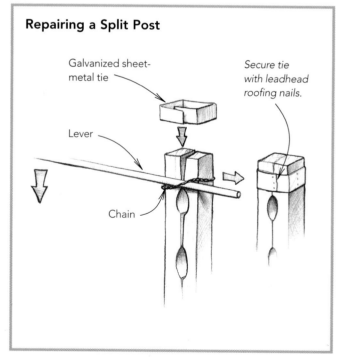

Galvanized sheet-metal tie

Secure tie with leadhead roofing nails.

Lever

Chain

Splicing New Wood onto a Damaged Post

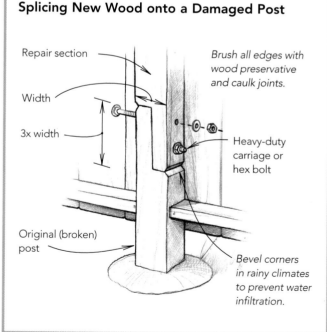

Repair section

Brush all edges with wood preservative and caulk joints.

Width

3x width

Heavy-duty carriage or hex bolt

Original (broken) post

Bevel corners in rainy climates to prevent water infiltration.

a strong lever (a crowbar or a hardwood 2x4 on edge works fine), as shown in the drawing at left on the facing page. Pull downward to close the gap and, while you maintain the pressure, have a helper band a strip of 2-in.-wide 18- or 20-gauge galvanized sheet metal around the top of the post and secure it with ring-shank, corrugated-steel roofing nails (known as "leadheads").

Occasionally, a post will be snapped off or damaged above ground at a point where enough good wood remains to splice in a repair section. Traditional timber framers employ any number of complex joints to splice together timbers that must handle simultaneous compression and bending loads. A simple half-lap splice is more than adequate to the requirements of a fence post. The only concern would be the possibility of water infiltration damaging the joint. This can be forestalled by caulking and, where appearance is not a problem and the local rains are persistent, by beveling the corners as shown in the drawing at right on the facing page. For strength, the splice should be about three times longer than the post width. Brush wood preservative onto all joint surfaces, even if using treated wood. Secure the lap with heavy-duty carriage or hex bolts (see pp. 111-112).

REPAIRING ROTTED POSTS

When posts rot (as opposed to being eaten by termites or tunneled by carpenter ants), the damage is usually confined to the buried portion. The

Like it or not, sometimes you have no choice but to pull a post. Posts that are driven into the ground or that lack a concrete footing can be pulled by hand, though it's a lot harder than you might think.

At first, it seems easy enough to do: The post will rock back and forth quite readily as you pull on it. Yet even though it rotates easily around the hole, the post will stubbornly refuse to rise more than 3 in. or 4 in. no matter how hard you lift on it. The suction of the damp earth and the barblike action of the flared post butt ensure that it will be difficult, if not impossible, to pull a post by hand without added leverage

You could use a tractor with a bucket loader to make light work of the job, but it really doesn't make sense to hire a tractor just to pull one or two posts. An alternative is to try the old-time trick of levering the post out of its hole. To gain the necessary mechanical advantage, first excavate around the base of the post, and then chain a stout lever (an 8-ft. hardwood 2x4 or a peeled-pole fence rail) as low on the post as possible. With the end of the lever bearing against a wooden pad, lifting up raises the post. Reset the chain and lever the post repeatedly until it can be lifted from the hole by hand.

The same technique can be used to turn a twisted post in its hole to bring its face back into harmony with the fence line.

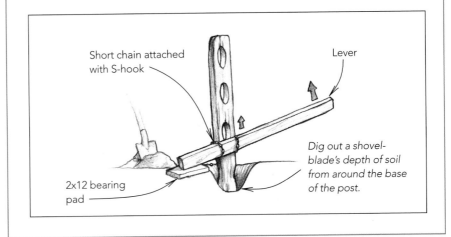

Short chain attached with S-hook

Lever

2x12 bearing pad

Dig out a shovel-blade's depth of soil from around the base of the post.

aboveground section remains intact. Repairing rotted posts is often a better idea than replacing them since you're likely to wreak considerable havoc on the stringers and infill boards when you try to detach them from the posts.

Begin by bracing the fence panels adjoining the defective post at right angles to the fence line to keep it from toppling over in the course of your renovations. Set blocks under the stringers on each side of the post. Dig out around the post down to its base.

The blocking will prevent the post from dropping and stressing the stringer joints once it's been sawn through and its decay-ridden lower end has been removed (along with any existing old concrete footing, which should bust apart fairly easily with a sledgehammer once the post is sawn through).

Saw the post off about 1 in. or 2 in. above grade. Check to be sure there's enough sound wood left at that point to make the repair feasible and that there's no sign of termite or ant infestation. If the rot extends more than a few inches further up into the core of the post, the only choice is to add more height to the repair post, trim the old post off at the point where the rot finally clears, and splice a filler piece onto the old post. Remove the broken concrete and any rotted wood from the posthole.

Cut a pressure-treated "sister post" of the same dimension as the old post. This sister post, whose aboveground length should be at least as long as its belowground depth but in no case less than 2½ ft., will serve to reinforce the existing post. Bevel the top of the sister post at a 45° angle so it sheds water and looks neater. Fill and tamp the bottom of the posthole with crushed stone as usual (see p. 160). Predrill the holes for the ½-in. H.D. (hot-dipped and heavy-duty) galvanized or satin-black-enamel-finish carriage bolts through the sister post first. The first bolt should be 6 in. to 8 in. above grade, with subsequent bolts every 6 in. to 8 in. apart and the last bolt no closer than 6 in. to the end of the piece. Select

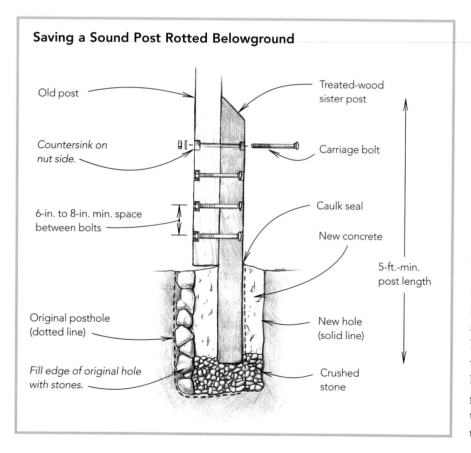

Saving a Sound Post Rotted Belowground

Old post

Treated-wood sister post

Countersink on nut side.

Carriage bolt

6-in. to 8-in. min. space between bolts

Caulk seal

New concrete

5-ft.-min. post length

Original posthole (dotted line)

New hole (solid line)

Fill edge of original hole with stones.

Crushed stone

bolts shorter than the full thickness of the two posts so that their heads can be countersunk on one or both sides as desired. (The round heads of carriage bolts are seldom objectionable and don't have to be countersunk.)

Position and tack the sister post to the existing post, and use its bolt holes as a guide for the long bit it will take to drill completely through both posts. (This kind of boring is more than the home-handyman drill can handle. You can rent the heavy-duty drill, but you'll probably need to buy the bit.)

Brush a coat of preservative on the abutting faces and on the bottom end of the old post. When the preservative is dry, run a bead of caulk down the mating edges of the posts and bolt the two posts together. Seal the top joint and coat the entire post with water sealer. Check the post for plumb and brace. To save concrete, line the edge of the old posthole with stones before pouring the new footing.

ADDING NEW POSTS

Sometimes there's no way around it: The stub of a sister post just looks too out of place to tolerate and the old post has to go. Assuming that the fence itself is or already has been trued to the line (see below) and that the only immediate concern is to provide a structural substitute for one or more of the posts, there's a less obtrusive and more architecturally consistent alternative: Instead of pulling the rotted post, leave it in place but take it off active duty, transferring its burden onto a new post set in the middle of the infill panel where it will almost look intentional (see the drawing on p. 220).

This approach does have some limitations. It's well suited for solid-board infills and for more open continuous-infill or picket-style fences, including arched and draped palings, so long as the top stringers or cap rails overrun the post tops or the posts otherwise end at or below the top of the infill. It could be used for board fences whose posts project above the infill, as long as the new posts are symmetrical and balanced with the old. With this approach, new posts can be added between existing posts of any fence without doing visual violence only if its overall composition does not depend on the serial progression of modular infill panels within which each post marks a strong caesura.

The only practical drawback to the repair is that you shouldn't add a new post to only one side of the newly nonstructural post. Adding new posts to either side allows the old one to keep the stringers and infill tied together, while the weight of the fence is shouldered by the new flanking posts. With this method, it's possible over time to gradually replace all the presently rotted and soon-to-rot untreated wood posts with treated lumber without tearing the fence apart or digging up the old posts.

However, even unburdened as they would be, leaving the old posts anchored in the ground is asking for trouble. Decay can progress farther up into the post and, as the below-grade portions rot away, water collecting in the old footing can cause uplift or settling. It makes more sense to cut the old posts off at ground level. This does, of course, add some lateral and vertical instability to the infill since the stringer connection to the old post now becomes a hinge, both horizontally and vertically. But unless the fence is unusually heavy, the stringers cantilevering off the new posts could conceivably carry their respective halves of the infill load without excessive strain. It would be wiser, though, to sister an on-flat 2x4 or even a 2x2 to the edge or underside of the original stringers to bind the hinge point shut.

Dig the new postholes, centered under the midspan of the infill panel. Notch both the replacement post and the existing stringer for an on-edge interface. With on-flat stringers, dado only the post face. Set, plumb, and brace the post, and then secure it to the stringers and the infill boards with screws and toenails as appropriate (see the new-post detail drawing on p. 220). Fill the posthole in the normal manner. Leave the diagonal braces in place for at least two or three days to let the concrete set up or the earth fill settle. Retamp and fill to grade after a rain.

Using a chainsaw or reciprocating saw, cut off the old posts just above ground level, making sure to block up under the stringers to each side of the post to relieve the load that could otherwise pinch the sawblade and cause

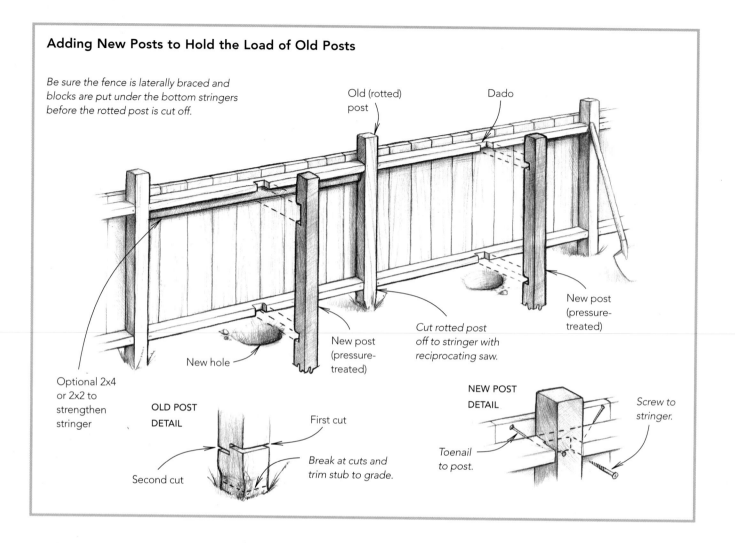

Adding New Posts to Hold the Load of Old Posts

Be sure the fence is laterally braced and blocks are put under the bottom stringers before the rotted post is cut off.

Old (rotted) post

Dado

New post (pressure-treated)

Optional 2x4 or 2x2 to strengthen stringer

New hole

New post (pressure-treated)

Cut rotted post off to stringer with reciprocating saw.

OLD POST DETAIL

First cut

Second cut

Break at cuts and trim stub to grade.

NEW POST DETAIL

Toenail to post.

Screw to stringer.

it to kick back with lethal force. To further protect against kickback, make two rough cuts, one above the other, from opposite sides of the post, stopping just short of each other. Break the hinge with a sharp hammer blow. With the strain thus relieved, make finish cuts flush to grade and to the bottom of the stringers.

FENCE FRAME AND PANEL REPAIRS

Most frame problems stem from post problems. At the same time, the stringers and the infill they carry do add significant racking resistance to the fence as a whole. When you attempt to true up that leaning fence, you'll learn just how stubborn a racked-out infill panel can be. A fence is a continuous structural membrane. If you push or pull on a post in the line of the fence (as opposed to at right angles to it), you push and pull against the dead load and the sprung joints and rigid connections of the entire fence, which is why it sometimes takes a lot more force to bring a fence back into line than you might think.

TRUING UP A FENCE

Truing up a fence is a lot like jacking up the sagging sills under an old house. When a large section has shifted, you can't realign it all at once. The job has to be done in small increments, often going back to the beginning with each lift or pull to reset the bracing for the next push. The braces that hold a leaning fence will be set and reset at the limit of whatever "give" each subsequent sequence of adjustments allows until the fence is gradually eased back into a straight and plumb line along its entire length.

There are a number of ways to pull a leaning fence back into line. For heavy fences that are seriously out of alignment, it really helps to use a come-along—so long as you've got something solid to hitch it to and enough cable to pull with. For something as pliable as a fence, the rear bumper of a pickup truck makes an ideal hitch.

Unless you have to true the fence within the plane of its run, however, you can usually move most fence posts simply by pushing on them by hand or by using some of the leverage tricks carpenters use to "line" a house wall frame. One such trick is to wedge a plank diagonally between the ground and the top of the post, pound the plank to push the post into plumb plus a "hair," and then tack it to the post to hold the angle. Alternatively, you can use a three-point triangular wall-lifting lever made by spiking a horizontal and a diagonal 2x4 to the bottom and top of the post, respectively, and spiking the end of the diagonal to the horizontal.

Whether you true the fence by hand or use a come-along, always begin by digging out the earth on the side of the post that faces the direction in which the fence must be moved. Removing the earth makes it easier to straighten leaning posts and avoids breaking

This leaning post-and-rail fence will need a lot of help to bring it back into line.

To paraphrase Robert Frost, "Something there is that doesn't love a fence." Plumbing and straightening a fence can be a constant chore.

concrete post footings. Don't remove any stringers.

Just as you would to line a house wall, plumb and brace the corner or terminal posts first, stretch a string between them on spacer blocks, and straighten the rest of the fence to this line. It's particularly important to take your time when straightening the fence parallel with the line of its run. Pulling on a come-along too fast can rip already stressed stringers off the posts. Once a joint is pulled open, the tangle of exposed toenails makes it even harder to close back up. Whenever practical, pushing is preferable to pulling.

Once the fence or a post has been pulled back onto the mark, holding it there can be another challenge. Automatic bracing is one advantage the come-along has over the other levering tools. With the come-along keeping the tension on the post or frame, you're free to set staked diagonal braces to

keep the post from relaxing back into its original repose when the tension is released. Otherwise, you'll need a helper to monitor the fence alignment with the spacer block or a level as you set the braces.

With the fence braced to the line, fence problems once again become post problems. The posts must be able to hold their new alignment after the temporary diagonal braces are removed. Digging out the old backfill and adding a crushed-stone base and repacking the hole with gravel instead of soft earth might be all that's required to set the fence firm. On the other hand, more radical measures may be called for. In stable soil, the sides of the hole can be undermined and a tangle of chicken wire spiked to the post base with galvanized nails to tie it to a freshly poured concrete collar. In unstable soils, strain plugs oriented to the line of the lean rather than to the line of the fence might also be needed.

REPAIRING ROTTED STRINGERS

Removing a damaged stringer will generally wreak havoc on the infill boards. Unlike separating a post from its infill, with stringers, there are simply too many nails to pry off without splitting and splintering, especially since aged infill is apt to be brittle and easily split in any case.

The least destructive way to remove a damaged or rotted stringer is to saw through the infill nails with a reciprocating saw from behind. Beginning at the top of the fence, tap the back of the infill board with a block to make room to insert the sawblade. After the top fasteners have been cut, hammer with the block against the infill at the middle stringer (if there is one) and saw through this line of nails as well. The nails don't have to be cut off perfectly flush. The stubs can be removed after the boards are off by pulling them out with a pair of pincers (or you can pound them flush).

It's not always necessary to cut through the nails to remove the boards. Sometimes small-headed stainless or galvanized wood shingle and siding nails can be pushed through the board from the face side with a pin punch or nail set. Once the upper two stringers have been released, the board itself provides the leverage to pry it away from the bottom stringer. Sometimes you may have to remove infill boards to gain access to both sides of a notched-in stringer, either to facilitate sawing

through, pulling toe nails, or even driving the stringer out sideways.

To repair a damaged section of stringer, cut out the unsound material with a handsaw and splice a new piece back in. If possible, make the cut in sound wood where the saw can slip in between a pair of pickets or fence boards. Otherwise, pry or push the

---

REPAIRING A PALING-STYLE STRINGER

The through mortises in the top and mid-stringers of dowel-style paling fences virtually guarantee the need for eventual stringer surgery. Boring round holes in replacement stock isn't hard, but unless you have a mortise-cutting attachment and a drill press, boring and chiseling out a long line of square mortises is an exacting and tedious chore.

A good shortcut, especially for diamond-pattern square mortises, is to make the replacement stringer in two halves, cutting the mortises into each with simple 45° angle cuts, and then glue and screw them together. The pales can be inserted after the stringer is installed or with it as part of a panel assembly when bottom and/or mid-stringers are also being replaced.

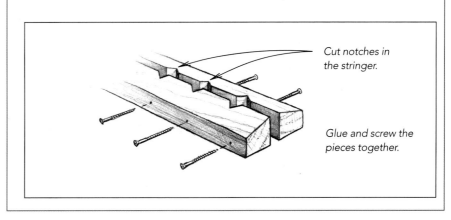

Cut notches in the stringer.

Glue and screw the pieces together.

boards away from the back of the stringer to give the tip of the sawblade enough "play" so that it doesn't jam against them. You could also make rough cuts with the reciprocating saw first and finish cuts with a handsaw afterwards.

Pry off the damaged piece of stringer. If the section is fairly long (that is, there are too many nails to pry), slice it into more manageable chunks and split them out with a chisel. Saw and chisel a half-lap rabbet into the remaining sound stringer. Cut a matching rabbet on the repair piece and bolt it to the stringer.

REPLACING ROTTED FENCE BOARDS

Removing fence boards is the same process as loosening them when facilitating stringer removal. The greatest problem with replacing fence boards is matching both the material and its finish, particularly when the boards are left to weather. The grain of the wood, either dressed or rough, and for rough boards, the characteristic signature of the sawblade, will become a visual eyesore if not matched against the original infill. Even the same wood won't have the same patina as the old. You can wait for it to age or use bleaching stains to "pre-age" it with variously successful results. If you have a can of the original stain or can identify it, test it on a scrap of new fence board first. You'll need to experiment until you come up with the closest match to the weathered original tone.

GATE REPAIRS

Gates most commonly sag because there is a problem with the hardware. The main cause of loose hinges is screws that are too short or screw holes that have become too sloppy. To tighten up a screw hole, hammer tapered splits (any hardwood or softwood will do) dipped in glue into the hole until it's packed solid. When the problem stems from hinges that are too small for the gate, replace them with beefier ones.

If hinges and screws are tight and strong, check the hingepost for lean along the drag line. Plumb the post, repairing or replacing its foundation if necessary. To reduce the likelihood of future recurrence, add a tension brace

A sagging gate is a common problem that can often be remedied simply by rehanging the hinges.

from the hingepost to the bottom of its adjoining line post (see the drawing at right).

By default, if the cause of the sag is neither the hardware nor the hingepost, it's the gate frame. True it up with a threaded turnbuckle tension rod (with flattened ends for screws). The tension rod works even with an existing wood brace that's no longer tightly held. Squaring the gate by installing a tension brace is easier than taking it apart and rebuilding the joints.

If a gate binds after a spell of damp weather but remains otherwise trouble-free, plane ¼ in. off its leading edge. When the opposite is true and the latch falls short of its catch in dry weather, reposition the latch or replace the bolt with a longer one. The hinge-shimming adjustments described in Chapter 7 can also be employed to shift the gate into a more favorable alignment with its post. Drill pilot holes through the shims so that the screws don't split them.

APPLYING COATINGS TO FENCES

Whether you're painting a section of fence that you've just replaced or coating a whole new fence, proper surface preparation is a key element in the success of the finish. It's important to clean the surface of all loose grime, dust, oil, and grease and to let the wood age before priming. Newly cut lumber contains salts and minerals that can

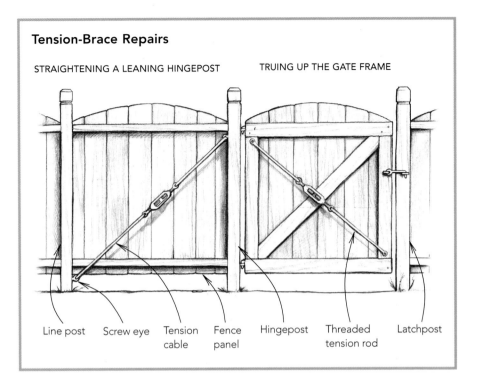

Tension-Brace Repairs

STRAIGHTENING A LEANING HINGEPOST

TRUING UP THE GATE FRAME

Line post | Screw eye | Tension cable | Fence panel | Hingepost | Threaded tension rod | Latchpost

destroy the paint coating from inside out as they leach from the wood fibers. Let green wood weather in place at least three weeks before attempting to paint it. Give it (and all fence boards) a coat of water repellent, wait at least another two days, and then prime it. A better strategy for seasoned green lumber is to buy and store it for several months before using it.

The best way to paint a fence is to do so before you build it. You may use more paint, but you'll save time and, more important, the hidden surfaces that would otherwise rot first will be protected. Complete coverage is especially important when brushing on

water-repellent or preservative coatings. These coatings, as well as semi-transparent stains and other penetrating finishes, are most efficiently applied by dipping, especially on rough wood (see the sidebar on p. 226).

Whether prestaining or painting in place, wait until the dew has evaporated before applying any stain or paint. The ideal temperature range for painting is between 60°F and 80°F; lower temperatures delay drying and can cause moisture to become trapped behind the paint film; higher temperatures cause the paint to dry out too fast, making a weak bond.

FENCE DIP

"Tom appeared on the sidewalk with a bucket of whitewash and a long-handled brush. He surveyed the fence, and all gladness left him and a deep melancholy settled down upon his spirit. Thirty yards of board fence nine feet high. Life to him seemed hollow and existence but a burden." (Mark Twain, *The Adventures of Tom Sawyer,* 1876)

If the thought of painting a fence by brush seems like a daunting and thankless task, consider coating the components before you build the fence. First, you'll need to construct a plywood trough large enough to hold a quantity of the longest fence components and line it with a double layer of 6-mil or 10-mil polyethylene sheeting. You could also dig the trough in the ground.

Fill the trough with enough sealer to cover one side of the fence boards and soak them in it for about a minute on each side. Lean the materials upright to dry or stack them in layers separated by narrow stringers. If dipping in stain, wipe the finished board off with a coarse towel before setting it aside to dry.

Gang-paint batches of fence material with primer and finish coats or solid-color stain by laying them across a pair of sawhorses set up on a ground cloth of plastic sheeting. Give the edges of each painted piece a quick pass of the brush or roller to remove the drips and runs that inevitably build up along them.

If possible, try to paint in the early morning hours, following the sun to work in the shade as much as possible.

On smooth surfaces, the coating should be applied as thinly as possible. Thick films will have difficulty keeping their grip as the wood beneath them contracts and expands with temperature changes. A thick coating is preferable on rough surfaces, where the paint can insinuate itself into the fibrous surface. Rough surfaces with thick coatings hold up better to extremes of temperature.

PAINTING TOOLS

Using the right tool for the job is just as important for painting a fence as it is for any other activity. A brush or roller is the best tool for a picket fence, whereas sprayers are ideal for painting board fences, rough or smooth. Some people recommend using a sprayer regardless of fence type, but on a picket fence most of the paint delivered by an airless sprayer will be blown between the pickets. To ensure better adhesion, always brush or roll on the primer coat rather than spraying it.

When buying a roller for fence painting, don't waste your money on

The important thing to keep in mind when painting a fence is that the end is in sight. Amen.

high-quality rollers. Fence painting is tough on rollers, especially on rough wood, so go with the cheap synthetics. I don't bother washing out rollers between uses. They'll keep for up to a day wrapped in aluminum foil or plastic wrap; after that, throw them away.

There are various types of sprayers suitable for coating a fence. You can use a garden-type sprayer to apply water-repellent and preservative sealers and other thin coatings, but these sprayers won't handle thick fence paint. The light-duty home-handyman-quality airless spray guns are okay for occasional use on small areas, but they aren't up to the demands of prolonged duty involved in painting a big fence. The 3-hp to 5-hp gas-powered airless sprayers that you can rent for about $45 a day can handle coatings as heavy as tar. Unlike conventional compressed-air sprayers where the paint is actually mixed with air, in "airless" sprayers the compressed air is directed into the paint tank to push the pressurized paint through the hose and the high-pressure nozzle that atomizes it into a cone-shaped spray. The result is considerably less overspray and less wasted paint.

Even so, less overspray can still mean a lot of paint drifting over your plants and lawn and onto your neighbor's new car. Spray paint only on calm days, and, for best results, hold the gun at a right angle to the fence, about 8 in. to 10 in. away from it. Protect any plants and other vulnerable nearby surfaces with plastic or heavy cotton dropcloths. Rest a sheet of cardboard against sawhorses to reduce overspray on picket fences. Follow the instructions and safety precautions that come with the sprayer, and wear long pants and a long-sleeved shirt, safety goggles, and a respirator.

BIBLIOGRAPHY

Allport, Susan. *Sermons in Stone: The Stone Walls of New England and New York.* New York: W. W. Norton, 1994.

Cleaveland, Henry W., William Backus, and Samuel D. Backus. *Village and Farm Cottages.* 1856. Reprint, Watkins Glen, N.Y.: American Life Foundation & Study Institute, 1976.

Collins, A. Frederick. *Keeping Your House in Repair.* New York: Appleton-Century Co., 1886.

Deane, Samuel. *The New England Farmer: or, Georgical Dictionary.* Boston: 1790.

Downing, Andrew Jackson. *Victorian Cottage Residences.* 1873. Reprint, New York: Dover Publications, 1981.

"Fences & Fence Posts of Colonial Times." From the *White Pine Series of Architectural Monographs*, Vol. VIII, #6, 1922.

Gillon, Edmund V., Jr. *Pictorial Archive of Early Illustrations and Views of American Architecture.* New York: Dover Publications, 1971.

Handlin, David P. *The American Home: Architecture and Society 1815–1915.* Boston: Little, Brown and Company, 1979.

Hill, Thomas E. *Hill's Manual of Business and Social Information.* Chicago: W. B. Conkey Company, 1921.

Jordan, Cora. *Neighbor Law: Fences, Trees, Boundaries and Noise.* Berkeley, Calif.: Nolo Press, 1991.

Lehner, Ernst and Johanna. *How They Saw the New World.* New York: Tudor Publishing, 1966.

Martin, George A. *Fences, Gates and Bridges: A Practical Manual.* 1887. Reprint, Brattleboro, Vt.: Alan C. Hood & Co., 1992.

Masury, Anne Rankin. *Portsmouth Fence Styles.* Unpublished master's thesis, University of Connecticut, 1979.

Palliser, Palliser & Co., Architects. *Palliser's New Cottage Homes.* New York: Palliser, Palliser & Co., 1887.

Russell, Howard S. *A Long, Deep Furrow: Three Centuries of Farming in New England.* Abridged edition, Hanover, N.H.: University Press of New England, 1982.

Scott, Frank J. *The Art of Beautifying Suburban Home Grounds of Small Extent.* New York: John B. Alden, 1886.

Sloane, Eric. *Sketches of America Past.* New York: Promontory Press, 1986.

Stilgoe, John R. *The Common Landscape of America, 1580–1845.* New Haven: Yale University Press, 1982.

Varro, Marcus Terentius. *On Agriculture.* Translated by William Davis Hooper, revised by Harrison Boyd Ash. Cambridge, Mass.: Harvard University Press, 1884.

Webb, Walter Prescott. *The Great Plains.* Boston: Ginn and Company, 1931.

INDEX

publisher JON MILLER

acquisitions editor JULIE TRELSTAD

assistant editor KAREN LILJEDAHL

editor PETER CHAPMAN

designer & layout artist AMY BERNARD

photographer, except where noted JAMES P. BLAIR

illustrator ROBERT LAPOINTE

indexer HARRIET HODGES

typeface HORLEY OLD STYLE

paper 80-LB. UTOPIA TWO

printer R. R. DONNELLEY, WILLARD, OHIO